All-Optical Noninvasive Delayed Feedback Control of Semiconductor Lasers

Sylvia Schikora

All-Optical Noninvasive Delayed Feedback Control of Semiconductor Lasers

 Springer Spektrum

Sylvia Schikora
Berlin, Germany

Dissertation Humboldt University of Berlin, 2012

Date of the Disputation: 17.4.2012
Referee: Prof. Dr. Fritz Henneberger
       Prof. Dr. Wolfgang Elsäßer
       PD Dr. sc. Serhiy Yanchuk

ISBN 978-3-658-01539-8   ISBN 978-3-658-01540-4 (eBook)
DOI 10.1007/978-3-658-01540-4

The Deutsche Nationalbibliothek lists this publication in the Deutsche Nationalbibliografie; detailed bibliographic data are available in the Internet at http://dnb.d-nb.de.

Library of Congress Control Number: 2013935111

Printed on acid-free paper

Springer Spektrum is a brand of Springer DE.
Springer DE is part of Springer Science+Business Media.
www.springer-spektrum.de

Für meine beiden treuen Gefährten,
3moe 3625 w19-A11-20
und 4moe 3068 - 477[1].

---

[1] Im folgenden als 'integrated tandem laser' und 'active feedback laser' bezeichnet.

Vorwort

Die Arbeit an dieser Dissertation stand rückblickend gleichzeitig unter einem guten und einem schlechten Stern. Während die Bedingungen und Ereignisse auf wissenschaftlicher Seite kaum hätten besser sein können und auch die Geburt meiner beiden Kinder Tabea und Paul während der Arbeit an dieser Dissertation jener keinen Abbruch tat, wurde das Fortkommen gegen Ende doch mehr und mehr durch unerfreuliche private Geschehnisse erschwert. Dass diese Dissertation doch noch zu einem äußerst erfolgreichen Abschluss fand, wurde nur ermöglicht durch die Hilfe meiner Familie, zuallererst meiner Schwester Kirstin, meiner Mutter Brigitte Schikora sowie meines Vaters Detlef Schikora. Unendlicher Dank geht hier insbesondere an meine Schwester, die beste Tante der Welt, die Millionen von Stunden die liebevolle Betreuung meiner Kinder übernahm, damit ich die Arbeit überhaupt zu Papier bringen sowie mich auf die Disputation vorbereiten konnte. Danke an meine Freundin Brigitte Baumgart, die ebenfalls viele, viele Stunden der Betreuung übernahm. Danke an Ralf Tönjes für die geduldigen Gespräche über Physik.

Mit Ede Wünsche hatte ich den besten Betreuer einer Promotion, den man sich wünschen kann. Er war Berater, Ermunterer, Motivierer, Lehrer, Begleiter und zugleich Doktoropa für meine beiden Zwerge. Vielen Dank, Ede! Danke an Herrn Henneberger für das super spannende Thema, das ich im Rahmen dieser Dissertation bearbeiten durfte. Ich fühle mich glücklich, in einer Gruppe gelernt zu haben, in der mit unendlicher Geduld reagiert wurde auf überraschende Schwangerschaft, Ausfallzeiten wegen Kinderbetreuung und so viel mehr kleinere und größere Probleme, vor denen man als Wissenschaftlerin mit kleinen Kindern urplötzlich steht. Vielen Dank auch an alle anderen Photoniker für die schöne Zeit, insbesondere an Oleg Ushakov, Achim Puls und Ilya Akimov für die fachliche Unterstützung! Vielen Dank an Evi Poblenz für die Unterstützung! Ohne Deine Hilfe bei der Erschaffung des Babyzimmers wäre alles sehr viel schwieriger gewesen!

Das unerlässliche Kernstück dieser Arbeit waren die vom Heinrich-Hertz-Institut Berlin entwickelten Mehrsektionslaser, deren Potential nicht genug gewürdigt worden ist. Bernd Sartorius, ohne Ihre Laser wäre diese Arbeit wohl nicht realisiert worden, vielen Dank! Ganz lieben Dank an Stefan Bauer, der für die ersten Wochen die Betreuung dieser Arbeit geleistet hat und mich in sämtliche nötigen experimentellen Geheimnisse eingeweiht hat. Dank Dir blieb mir die schwierige Anfangsphase und Orientierungslosigkeit erspart. Danke an Olaf Brox, für viele Tipps und aufmunternde Gespräche. Vielen Dank auch an Matthias Biletzke und an Jochen für ihre Unterstützung und an die anderen Mitarbeiter vom HHI für die schöne Zeit dort.

Dr. Güther vom FBH Berlin hat mich umfangreich in der Frage der zu verwendenden Linsen beraten und ausführlich dazu gerechnet, danke für diese Unterstützung! Vielen Dank an die Mitarbeiter des Labors für Kohärenzoptik am Institut für Physik, für ihre kostbaren Ratschläge zum geeigneten Stabilisieren des Aufbaus. Danke an die Kollegen vom IKZ Berlin für die Bereitstellung maßgeschneiderter Siliziumscheiben. Ohne Dr. Dudeck and Herrn Rohde würde der experimentelle Aufbau nicht existieren, vielen Dank vor allem für die Bearbeitung der diversen Sonderwünsche!

Während dieses ersten Kennenlernens der Wissenschaft hatte ich das Glück, auf viele wunderbare Kollegen zu treffen. Für viele aufmunternde und interessante Gespräche möchte mich bei Uwe Bandelow, Bernd Krauskopf, Ingo Fischer, Thomas Erneux, Jan Sieber und Ralf Tönjes bedanken. Danke an Vasile Tronciu, Mindaugas Radziunas, Serhiy Yanchuk und an die anderen Kollegen vom WIAS für die Ratschläge in theoretischen Fragen! Danke an Eckehard Schöll, Bernold Fiedler, Philipp Hövel, Valentin Flunkert, Thomas Dahms für die Gespräche und die theoretische Unterstützung.

Danke an Tabea und Paul, ihr habt mir die nötige Kraft gegeben, durch euer Sein und Wunderbarsein.

Sylvia Schikora

Abstract

The stabilization of unstable states hidden in the dynamics of a system is interesting by itself, but also has the potential for useful applications. The control of unstable periodic states embedded in chaotic dynamics - referred to as 'control of chaos' - has received much attention in the last years.

In this work, a well-known control method called delayed feedback control is applied for the first time entirely in the all-optical domain. A multisection semiconductor laser is combined with an external Fabry-Perot interferometer, which provides a direct reflection of the optical output back into the laser. The interferometer creates a phase-tunable superposition of the laser signal. By feeding back such a control signal, the laser under certain conditions operates in an otherwise unstable periodic state with a period equal to the time delay. The control is noninvasive, because the reflected signal tends to zero when the target state is reached. In proof-of-concept experiments, all-optical control is applied successfully to different dynamical regimes:

- unstable steady states

- unstable periodic selfpulsations

- control of chaos

- unstable torsionfree periodic selfpulsations.

In the optical setup, a new degree of freedom comes into play - the wave phase. It makes all-optical delayed feedback control conceptually different from the standard approach. By tuning the optical phase, a switching between oppositional regimes like stabilization and destabilization is possible. A specific aspect is proved experimentally as well: the proposed potential of the phase-dependent setup to overcome limitations inherent to standard delayed feedback control.

The experiments are performed with a practical laser device. They operate in the picosecond domain, and thus on timescales much shorter than in any previous study. Being limited only by the speed of light, the method is in principle applicable to even faster dynamics.

Zusammenfassung

Das Stabilisieren instabiler, in der Dynamik eines Systems verborgener Zustände ist interessant an sich, birgt aber auch praktische Anwendungsmöglichkeiten. Insbesondere der 'Kontrolle von Chaos' - der Kontrolle instabiler regulärer Zustände innerhalb eines chaotischen Attraktors - wurde in den letzten Jahren viel Aufmerksamkeit gewidmet.

In dieser Arbeit wird eine bekannte Methode, zeitverzögerte Rückkopplungskontrolle, zum ersten Mal rein-optisch umgesetzt. Ein Mehrsektions-Halbleiterlaser wird mit einem externen Fabry-Perot-Interferometer kombiniert, welches die Emission direkt in den Laser zurück reflektiert. Das Interferometer erzeugt eine phasenabstimmbare Überlagerung des Lasersignals. Unter bestimmten Umständen führt die Rückkopplung eines solchen Kontrollsignals dazu, dass der Laser in einem ansonsten instabilen periodischen Zustand mit einer Periode gleich der Verzögerungszeit arbeitet. Die Kontrolle ist nichtinvasiv, da das reflektierte Signal verschwindet, sobald der Zielzustand erreicht ist. In 'proof-of-concept' Experimenten wird die Methode erfolgreich auf verschiedene dynamische Regime angewendet:

- instabile stationäre Zustände

- instabile periodische Selbstpulsationen

- Chaoskontrolle

- instabile torsionsfreie periodische Selbstpulsationen.

Der optische Aufbau bringt einen neuen Freiheitsgrad ins Spiel - die Wellenphase. Daher ist optische Kontrolle mittels zeitverzögerter Rückkopplung konzeptionell verschieden vom nichtoptischen Ansatz. Durch Anpassung der optischen Phase ist ein Schalten zwischen so gegensätzlichen Regimen wie z.B. Stabilisierung und Destabilisierung möglich. Das vorhergesagte Potential dieses phasenempfindlichen Aufbaus, Beschränkungen der nichtoptischen Kontrollmethode zu überwinden, wird experimentell bestätigt.

Die Experimente wurden mit einem praxisnahen Bauteil durchgeführt. Sie arbeiten im Pikosekundenbereich, und damit auf kürzeren Zeitskalen als in allen früheren Studien. Da die Methode nur durch die Lichtgeschwindigkeit beschränkt ist, kann sie im Prinzip auf noch schnellere Zeitskalen ausgedehnt werden.

Contents

Abbreviations   **XIX**

1 Introduction   **1**
    1.1   Chaos Control   2
    1.2   Delayed Feedback Control (DFC)   3
    1.3   All-Optical Delayed Feedback Control   5
    1.4   Content and Structure of This Thesis   7

2 All-Optical Control Setup   **9**
    2.1   Multisection Distributed Feedback Lasers   10
    2.2   Experimental Control Setup   15
    2.3   Numerical Modelling of the Control Setup   26
    2.4   Summary and Comparison with Literature   29

3 Stable States with Resonant Fabry-Perot Feedback   **31**
    3.1   Stationary States   31
    3.2   Controlling the Coherence of Oscillations   37
    3.3   Summary   41

4 Control of an Unstable Stationary State   **43**
    4.1   Stabilization of an Unstable Focus in Experiment   45
    4.2   Investigation of Control Parameter Space   48
    4.3   Summary   53

5 Control of Unstable Selfpulsations   **55**
    5.1   Experimental Optical DFC of a Periodic State   56
    5.2   Domains of Control   60
    5.3   Summary   63

6 Controlling Chaos   **65**
    6.1   Chaos Control in Experiment   66
    6.2   Domains of Control   71
    6.3   Summary   74

7   Control of a Torsionfree Orbit   75
      7.1   The Odd-Number Limitation in DFC   75
      7.2   Subcritical Hopf Bifurcation in the Freerunning Laser   79
      7.3   Recovering Hysteresis by Optical Control   82
      7.4   Control of the Torsionfree UPO in Simulations   85
      7.5   Experimental Control of a Torsionfree Orbit   87
      7.6   Summary   91

8   Conclusion   93

Appendix   95

A   Constraints of Delayed Feedback Control   97
      A.1   The Main Control Parameters   97
      A.2   The Odd-Number Limitation   98

B   The Amplitude Reflected by a Fabry-Perot Interferometer   99
      B.1   The Control Signal in Optical DFC   99
      B.2   The Intensity Transmitted by the FP   102

C   Noninvasive Control and Noise in Literature   103

D   Publications and Presentations   105

Bibliography   107

List of Figures

1.1   Schematics of a closed-loop control process   2
1.2   Unstable periodic orbit (UPO) in chaos   3
1.3   Sketch of all-optical DFC setup   6
1.4   Comparison: periodic signal and modulated wave   7

2.1   Structure of the integrated tandem laser   10
2.2   Photograph of the integrated tandem laser chip   10
2.3   Wavelength shift of a semiconductor DFB laser   11
2.4   Main dynamical regimes of the integrated tandem laser   12
2.5   Dynamical regimes in the plane of the laser currents   13
2.6   Route to chaos in the integrated tandem laser   13
2.7   Structure of the active feedback laser   15
2.8   Dynamical regimes in the active feedback laser   16
2.9   Experimental setup for optical DFC   17
2.10 Optical light extraction   19
2.11 FP transmission and reflectivity spectrum   20
2.12 Adjustment of a selfpulsating optical signal to a Fabry-Perot cavity   22
2.13 Photograph of the experimental control setup   25
2.14 Control setup in the numerical simulations   26

3.1   Figure from [TWWR06]. External cavity modes (ECM) in the (ω, N) plane   32
3.2   Measured transmission spectrum of the 26 GHz etalon   33
3.3   Tuning of latency phase φ after the FP is coupled resonantly to the cw emitting laser   34
3.4   Bifurcation diagram in the ($|E|$, φ) plane from [TWWR06]   35
3.5   Control of damped relaxation oscillations vs latency phase   36
3.6   Control of damped RO in dependence on the delay τ   37
3.7   Periodic orbit in phase space   38
3.8   Coherence improvement by resonant FP feedback   39
3.9   Coherence control in dependence on the latency phase   40

4.1   Unstable stationary states in 2-dim phase space   44
4.2   Supercritical Hopf bifurcation in 2-dim phase space   45
4.3   Supercritical Hopf bifurcation in the ITL   46
4.4   Control of unstable focus   47
4.5   Control of unstable focus versus the latency phase φ   48
4.6   Optical DFC in numerical simulations   49
4.7   Control of an unstable focus in the optical domain   50
4.8   Transient of the emitted output when switching control on and off   50
4.9   Control domains in (φ, K) plane for different latency times τ_l . . .   52
4.10 Control domains in (φ, τ_l) plane for $K = 0.05$ 52

5.1   Period doubling bifurcation   55
5.2   Period doubling route to chaos in the integrated tandem laser . . .   57
5.3   Control of the period-1 pulsation   58
5.4   Phase-dependent control of the period-1 pulsation   59
5.5   Control of the period-1 UPO versus φ   60
5.6   Phase-dependence of the emission wavelength under control . . .   61
5.7   Control of the period-1 selfpulsation in simulations   62

6.1   All-optical control of chaos   67
6.2   Control of chaos (optical spectrum)   68
6.3   Effect of the latency phase φ on optical chaos control   69
6.4   Control of chaos: Check of noninvasivity   70
6.5   Chaos control: simulations   72

7.1   UPO with and without torsion in phase space   77
7.2   Subcritical Hopf bifurcation   80
7.3   Controlling bistability at the subcritical Hopf bifurcation   81
7.4   Control of focus against noise   83
7.5   Simulations: Control of UPO at a subcritical Hopf bifurcation . .   84
7.6   Simulations: Control of UPO at a subcritical Hopf bifurcation ver-
      sus phase   86
7.7   Control of UPO in dependence on φ   88
7.8   Control experiments at the subcritical Hopf bifurcation   89
7.9   Details of the controlled UPO at the subcritical Hopf bifurcation .   89

B.1   Sketch of FP setup   99

List of Tables

2.1   Simulation parameters   29

Abbreviations

OGY	Ott, Grebogi, and Yorke
UPO	unstable periodic orbit
DFC	delayed feedback control
FP	Fabry-Perot
MSL	multisection semiconductor laser
DFB	distributed feedback laser
ITL	integrated tandem laser
AFL	amplified feedback laser
cw	continuous wave
FWHM	full width at half maximum
IR	infrared
ESA	electrical spectrum analyzer
EDFA	erbium doped fiber amplifier
TW	traveling wave
RO	relaxation oscillations
ECM	external cavity modes
SLM	solitary laser mode
SP	selfpulsation
SN	saddle-node bifurcation

1 Introduction

Feedback and control are closely related concepts known within all natural sciences, engineering, as well as the social and life sciences. Any self-regulating process, whether a natural or an artificial one, is involving some type of feedback. 'Feedback' refers to a mechanism, in which the output signal of a system is partly returned to the system as input. The effect of feedback can be twofold. Negative feedback acts decelerating and tends to maintain stability. In contrast, positive feedback acts accelerating and induces rapid changes. In natural systems, both types of feedback can be found, with the negative feedback being more common. Normally, a natural process requires the parameters of the system to stay within a certain range. An example is the temperature regulation, e.g. in organisms or in ecosystems, where the system needs the stability of this parameter simply to exist. To this end, a feedback loop is created that controls the parameter.

In general, 'control' means that the behavior of a system is steered deliberately to achieve a desired performance. Researchers adopted the principle from nature very long ago. Earliest examples for control of dynamical systems date back to antiquity[1]. The formal analysis in control theory began with a classical paper by J.C. Maxwell from 1868 [Max68], where he considered, amongst others, the steam engine by James Watt.

Today control theory represents a well-established interdisciplinary subfield of engineering and mathematics. Applications put forth by the age-long research in this field are indispensable in everyday life, ranging from refrigerators over water closets to the steering of spacecrafts. Most of the known control schemes rely on feedback, and are closed-loop controllers. The functionality of a closed-loop controller is as follows (Fig. 1.1): a sensor monitors the output of the system, passes the data to the controller, which compares the output with a reference state and generates an appropriate control signal (or 'control force') which is fed back to the system. This way, a dynamic compensation for disturbances can be achieved.

---

[1]One example for control mechanisms used in that time are float valves controlling the flow in greek and roman water clocks.

1.1 Chaos Control

Though in the meantime it had become a central field in applied mathematics and engineering science, control was kind of rediscovered for the physical society in 1990 in connection with the rather young phenomenon 'chaos' [LY75]. Ott, Grebogi and Yorke (OGY) [OGY90] pointed out the possibility to 'control' chaos[2] by stabilizing one of the infinitely many unstable periodic orbits (UPOs) embedded in chaotic attractors (Fig. 1.2) [ER85]. The controller suggested by OGY compares the actual system state with an (unstable) reference state by means of continuous computations. Due to the ergodicity of chaos [LY75], the trajectory of a chaotic system approaches every embedded UPO arbitrarily close over time. Whenever the actual state differs from the reference state, but is close in the sense that the computed deviation is below a certain upper bound, the controller applies a suitable correction to a control parameter of the system. As a result, the trajectory of the otherwise chaotic system moves along an UPO that exists within chaos.

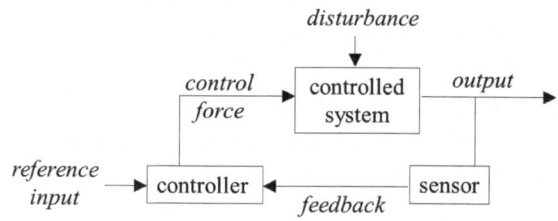

Figure 1.1: *Schematics of a closed-loop control process.*

This finding completely changed the attitude of scientists towards chaos, raising the former undesired phenomenon to an attractive object of research. One reason is that knowledge about UPOs in a chaotic attractor may yield additional information about the attractor itself. For instance, the topological entropy or the Hausdorff dimension can be extracted [ACacE+87]. Further, a characterization of the topological properties of the attractor is enabled [PFA+92]. Control of chaos is especially interesting from the engineering point of view. A chaotic system bears the possibility to control different types of motion (corresponding to different UPO's) in one and the same system with small control power. Owing to the extreme sensitivity of chaotic systems to external perturbations, the necessary control forces

---

[2]Misleading, the process is called 'control of chaos', but actually refers to the turning of a chaotic signal into a regular signal.

are very small[3]. The OGY method and its modifications have been successfully applied in experiments, for example, with a chaotic flow in a thermal convection loop [SWB91], with a diode resonator [Hun91] or with a solid-state laser system [RMM+92]. However, the OGY scheme requires detailed knowledge of the target state and is limited in speed due to the necessary computations.

Figure 1.2: *Unstable periodic orbit (UPO) in chaos. The chaotic attractor (black line) of the investigated tandem laser device (see Chapter 2) and one of its UPO's (thick gray line) in two-dimensional phase space.*

1.2 Delayed Feedback Control (DFC)

After the seminal paper by Ott, Grebogi and Yorke, a flood of proposals for further chaos control methods followed. Among them, a method called 'delayed feedback control' (DFC) or 'time-delayed autosynchronization' is outstanding because of its simplicity. It has been proposed and numerically confirmed by Pyragas in 1992 [Pyr92], followed by an experimental verification in 1993 [PT93]. The Pyragas scheme acts directly on a dynamic variable of the system, rather than on a system control parameter like in the OGY method. Negative feedback of a time-delayed difference of this dynamic variable is applied to the system. This way, the comparison with a reference state, which would require complete knowledge about the state, is avoided. Instead, the actual value of the dynamic variable is continuously compared with a value delayed by a certain delay time τ in the past. This approach yields a self-adjusting controller that needs no on line data processing or detailed

---

[3] Already in 1985, i.e. previous to the invention of chaos control, NASA engineers made use of this property. They steered a spacecraft with an insufficient amount of fuel left over 50 million miles for the first comet encounter in history [FMD85].

knowledge about the underlying dynamics. DFC can therefore be applied even to extremely fast systems where real-time computations are not possible, or to those nonlinear systems which cannot be modelled theoretically.

The experimental application of DFC is very straightforward: one only needs to measure an appropriate variable $s(t)$ of the dynamic system and generate a difference of this output variable and of a time-delayed version of it. The control signal

$$F(t) = -K(s(t) - s(t - \tau)), \qquad (1.1)$$

with negative sign, convenient feedback strength K and certain delay time τ, is acted back to the system. Given that an UPO with a period T equal to τ exists, this orbit is locally stabilized by the feedback. The system approaches the UPO in a self-adjusting way[4]. Thereby, the applied control force (1.1) decreases and finally vanishes as the UPO is reached, because then $s(t) = s(t - \tau)$. Whenever perturbations cause the system to leave the UPO, a small feedback arises to direct the system back. That reflects an important feature of DFC: the control force (1.1) is zero as long as the system remains on the desired orbit, the controller acts 'noninvasive' with respect to the stabilized state. The stabilized orbit thus represents a genuine state of the dynamical system, only its stability is changed[5]. In experimental realizations, the smallness of the remaining feedback signal under control is only limited by the noise level in the system.

There are different variations of DFC. One is particularly important in context with this work: extended delayed feedback control [SSG94]. Here, multiple delays are included in the special form

$$F(t) = -K \sum_{n=0}^{\infty} R^n [s(t_n) - s(t_n - \tau)], \qquad (1.2)$$

with $t_n = t - n\tau$. An additional control parameter appears: the memory parameter R. It steers the impact of states further in the past within the control force. The original Pyragas scheme (1.1) is contained in (1.2) as special case $R = 0$. Already when suggesting this scheme in [SSG94] the authors pointed out its inherent potential to be realized by purely optical means, as Eq. (1.2) corresponds to the signal reflected by a plane Fabry-Perot interferometer. Nevertheless, this matter thereafter has not been followed up further, a deeper theoretical analysis or experimental approaches are not reported.

---

[4]To be precise, the dynamic system can approach the UPO only when started in the basin of attraction of the orbit under control.
[5]The same holds for an orbit stabilized by the OGY method.

Since its invention, DFC has found a deep theoretical foundation as well as numerous applications, see, e.g. the reviews [BGL+00, Pyr06, SS08]. The experimental applications mostly employ electronic feedback loops. Some characteristics of DFC important for this thesis are summarized in Appendix A.

Involving no numerically expensive computations, DFC is capable of controlling systems with fast dynamics still in real-time mode. This is particularly important for semiconductor lasers. Technical progress goes towards increasingly higher speed of operation. Multisection semiconductor lasers have been operated at tens of GHz [BSS04, Slo05] and the THz range is in view [AMKK+07, HHD+10]. These picosecond time-scales are too short even for fast electronic control circuits - all-optical control setups are required. In optical communication systems semiconductor lasers nowadays play a key role. Especially multisection semiconductor lasers have opened up new ways in high speed optical information processing [BBK+04, WBK+05]. Due to the inherent nonlinearities, they can emit not only continuous wave but also periodic pulse trains or chaotic light streams. In optical communication, extremely stable periodic pulsations on timescales of tens of GHz are of interest. Nevertheless, the importance of chaos is also increasing, it either has to be avoided [RDLG07, LRL+07], or it is exploited, like, for instance, in the developing chaos communication [Cry02, ASL+05, PRW+06, UAI+08, KMH09]. Therefore, it is desirable to realize an ultrafast control setup for semiconductor lasers which works entirely in the optical domain where the velocity of light sets the ultimate speed limit.

1.3 All-Optical Delayed Feedback Control

This work aims at the experimental realization of all-optical ultrafast DFC of multisection semiconductor lasers and the study of its specific properties. Fig. 1.3 depicts the chosen basic configuration with feedback from a Fabry-Perot resonator (FP).

This configuration [SSG94, SBGS97, SH96] and a different one exploiting feedback from a Michelson interferometer [LH94] have been proposed already more than a decade ago. The Michelson configuration is the optical version of the original Pyragas method (Eq. (1.1)), whereas the FP setup represents the extended version of DFC corresponding to Eq. (1.2). In both cases, the variable s is given by the optical field amplitude E emitted by the laser. In the chosen FP configuration (Fig. 1.3), the laser emission with amplitude $E(t)$ is forwarded into a plane FP cavity with mirror reflectivity R. The mirror separation defines a cavity roundtrip time τ, which realizes the required time delay. A part $E_b(t)$ of the optical am-

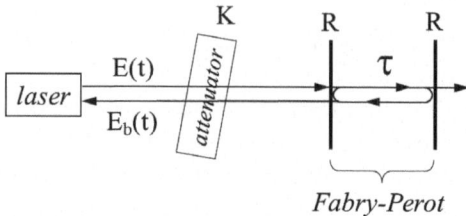

Figure 1.3: *Sketch of all-optical DFC setup. A laser receives feedback from a plane FP resonator with mirror reflectivity R and internal delay τ. An attenuator adjusts the feedback strength K.*

plitude is reflected at the cavity mirrors and reinjected into the laser. This way, a control signal of the form $E(t) - E(t - \tau)$ can indeed be constructed under appropriate conditions. Compared to DFC in the electrical domain, the optical phase becomes important and its role must be thoroughly analyzed.

Though proposed early, this configuration has not been realized in experiment so far. One reason for missing experimental implementations can be a difficulty that lies in the relatively long time scales on which the chaotic behavior of most lasers proceeds. They require FP cavity lengths that are either completely impractical, or where the efforts to stabilize the cavity are so extensive that methods relying on optoelectronic feedback are much better suited. For multisection semiconductor lasers the situation is just reversed: optoelectronic control is no longer practical, while the dimensions of suitable external FP cavities shrink to the few-centimeter range. FP cavities of this size can be easily combined with the laser in a robust setup. A future option is even integration of laser and FP cavity in a single device.

This book systematically presents my results on the realization of all-optical control of multisection semiconductor lasers in various dynamical states of different complexity. The main results are published in [SHW⁺06, SWH08, WSH08, SWH11, DFH⁺10]. Great importance is attached to the study of a specific feature of optical DFC which has been overlooked in the original proposals in [SSG94, SBGS97]. This specific results from the fact that the pulsating emission of a laser represents a modulated optical wave. Fig. 1.4 demonstrates the difference between a simple periodic signal and a periodically modulated carrier wave. Eq. (1.2) for non-optical DFC aims at the stabilization of simple τ-periodic orbits via a signal $s(t)$ with the same period (Fig. 1.4 (a)). In optical DFC of τ-periodic pulsations, the optical field is a τ-periodic modulated optical wave (Fig. 1.4 (b)). The consequence for optical feedback is that here the phase shift of the carrier wave along

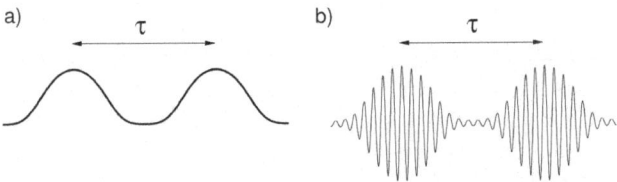

Figure 1.4: *Schematic comparison of a simple periodic signal (a) and an accordingly periodically modulated carrier wave (b).*

the optical path additionally plays a role. Therefore, Eq. (1.2) is actually only half the truth and must be completed by additional phase contributions, leading to Eq. (2.1) below for the phase-dependent control signal in optical DFC. On the one hand, this represents an experimental challenge, because all relative positions of the laser and the mirrors of the resonator have to be adjusted with subwavelength precision and held stable. On the other hand, this leads to a new control parameter and thus to new possibilities of DFC. By tuning the optical phase - the new control parameter - the sign of the control signal F can be reversed, and an easy switching between negative feedback and positive feedback to the system is possible. Thus, both oppositional main feedback regimes - destabilization/enhancement and stabilization/deceleration - can be experimentally studied by just tuning one parameter.

1.4 Content and Structure of This Thesis

This thesis is organized as follows. Chapter 2 describes the basic design of the experimental control setup, including the utilized laser devices, and the model used for theoretical analysis. The experiments were accompanied and partly preceded by numerical experiments with a traveling wave model on a full-device level. This has been enabled by a powerful simulation tool developed by M. Radziunas (http://www.wias-berlin.de/software/ldsl/). The theoretical results important for this work are not summarized in a separate theoretical Chapter, but will be presented individually where they are relevant.

Chapters 3 to 7 present applications of the all-optical control method on different scenarios. They are arranged according to an increasing level of complexity of the scenarios and, thus, increasing experimental challenge. In each experiment, the closeness of the point of operation to a specific bifurcation enables defined conditions. Before turning to the stabilization of unstable states, Chapter 3 deals with the application of all-optical DFC on stable states. The subsequent chapters study the stabilization of unstable steady states (Ch. 4), unstable periodic orbits close to a bifurcation (Ch. 5), and within chaos (Ch. 6). Finally, a specific type of unstable periodic orbits is studied, which until very recently [FFG$^+$07] has been believed to be non-stabilizable by DFC (Ch. 7). The last Chapter summarizes the achievements. Three appendices are concerned with more specific aspects. Appendix A summarizes the main constraints of the delayed feedback control method, in Appendix B the optical amplitude reflected by a FP interferometer is derived, and Appendix C reviews how the problem of 'noninvasive' control under noise is discussed in literature.

2 All-Optical Control Setup

The aim of the present thesis is the developing and realization of an experimental setup qualified for DFC of semiconductor lasers by purely optical means, and the demonstration of all-optical control of chaos in a semiconductor laser. Normally, semiconductor lasers do not show chaotic dynamics by themselves. However, it is possible to generate chaos in semiconductor lasers. This can be achieved, for instance, by addition of (optical) time-delayed feedback [LK80], or by the mutual delay-coupling of two semiconductor lasers [WBK+05]. Both principles can be realized in a compact way by a multisection semiconductor laser (MSL). In the first case, additional delayed feedback sections complement the laser, while in the second case two lasers are delay-coupled on the same chip. In this work, two types of MSL are studied which realize the above principles in the limit of ultra-short delay [UBB+04] (Fig. 2.1 and Fig. 2.7). Their rich dynamics make these devices ideal candidates for the planned experiments. Sec. 2.1 introduces the two MSL devices studied throughout this work and their specific dynamics.

Subsequently, the experimental conditions for optical DFC with these lasers are discussed. Normally, the MSL devices are available in a packed form as plug-and-play module. Unfortunately, these modules are of no use here because they impede a reasonably short latency time, that is, the travelling time of the signal between laser and controller. Thus, a setup has to be created which provides for the stable operation of bare MSL chips, for the extraction of their optical output and its analysis, as well as for the coupling of a FP interferometer to one laser facet. The experimental implementation is described in Sec. 2.2.

The experiments are accompanied by numerical simulations of the control setup on a full-device level. In the Photonics group of Humboldt University, where the present thesis has been worked out, there is wide experience with the numerical study of semiconductor lasers. A strong simulation tool has proved to reproduce MSL dynamics realistically (http://www.wias-berlin.de/software/ldsl/). This enables a preliminar testing of suitable ranges for the relevant control parameters, and allows for further explorations of the control parameter space beyond the points of operation chosen in the experiments. The used traveling wave model [ZSFYMC94, HMM96, SM06, Rad06] and the numerical setup are introduced in Sec. 2.3.

2.1 Multisection Distributed Feedback Lasers

The studied MSL devices are 1550 nm emitting distributed feedback (DFB) lasers basing on a bulk InGaAsP heterostructure. They were developed and fabricated at the Fraunhofer Institute Heinrich-Hertz-Institut Berlin by Bernd Sartorius and his coworkers. Their operation principles are described in [Bro05, Bau04, BBK$^+$04]. The two devices employed in this thesis are an integrated tandem laser (ITL) and an active feedback laser (AFL). The ITL consists of two mutually coupled distributed feedback lasers, while the AFL comprises a DFB laser and integrated ultrashort optical feedback consisting of two parts: a passive waveguide and an amplifying section.

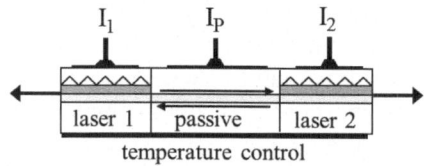

Figure 2.1: *Structure of integrated tandem laser (ITL), longitudinal cut.*

Integrated Tandem Laser (ITL)

The ITL unites two single-mode DFB lasers which are grown together on one chip. Each DFB laser has a length of 230 μm and emits in the 1550 nm wavelength range used in telecommunications. The emission of a solitary DFB laser of this type is stationary. The wavelength of the stationary emission can be tuned by the

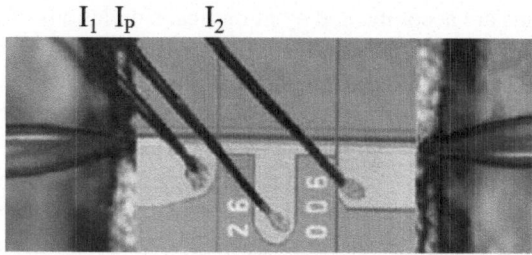

Figure 2.2: *Photograph of the ITL chip.*

injected current (Fig. 2.3). An increase of injection current leads to an increase of the wavelength due to a temperature induced change of the refraction index in the waveguide. Beyond that, a solitary DFB laser does not exhibit dynamics which are of interest in this work.

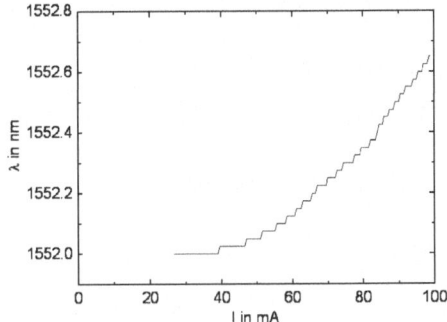

Figure 2.3: *Wavelength shift versus injection current for a semiconductor distributed feedback laser.*

In the ITL, two lasers of this type are mutually coupled by a 500 μm passive waveguide section (Fig. 2.1). The internal coupling strength K_c between both lasers is rather large: $K_c^2 = 0.5$. In order to avoid further interferences, the two laser facets are antireflection coated and the reflectivities at the interfaces are also negligible.

The three sections of the ITL can be independently biased by injection currents I_1, I_P and I_2 (see the three parallel wires in Fig. 2.2). The effect of a current variation depends on the type of the section. Current injection in the laser sections allows for a tuning of the emission wavelength with a sensitivity of 20 nm/A, similar to the solitary laser. Thus, in the ITL the optical frequencies of the two coupled lasers are tunable independently from each other. Current injection to the passive section changes the refraction index via free-carrier transitions. This way, changing the current I_P affects the internal coupling phase between the lasers. For this reason, I_P will be referred to in the following as 'phase current'. In all experiments, the phase current I_P serves as the main bifurcation parameter.

Depending on its specific parameters as e.g. the lengths of the sections or the wavelength detuning between the two lasers the ITL can exhibit more or less large regions of chaotic emission. Among various similar devices, the ITL device showing the largest chaotic regions in phasespace and the best similarity between the two DFB sections is selected. In this device, the threshold currents of the two laser

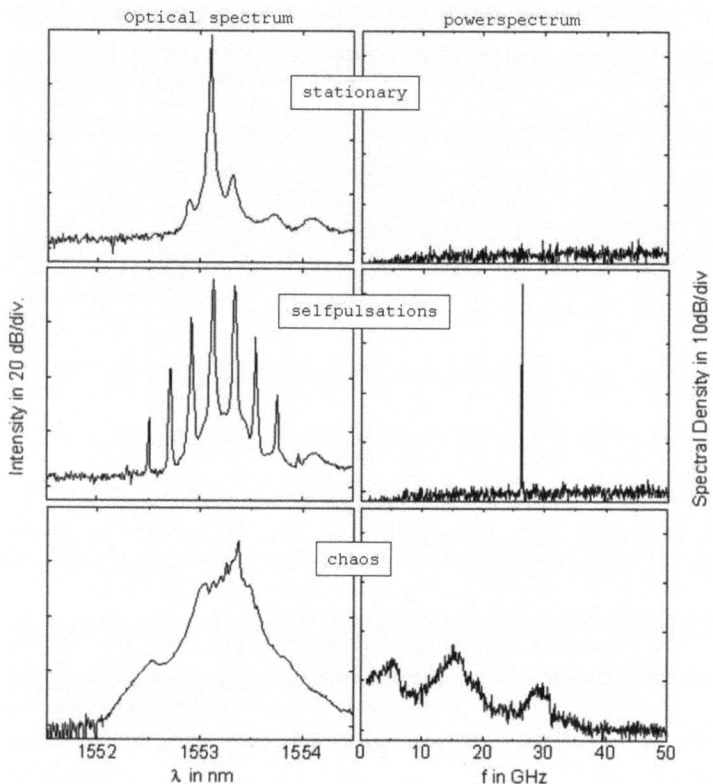

Figure 2.4: *Main dynamical regimes of the ITL. Left: optical spectra of the output, right: powerspectra. From top to bottom: cw emission, selfpulsations, chaos. Note the log scales.*

sections are 12 and 17 mA, respectively. The two lasers are slightly detuned in wavelength by about $\Delta\lambda = 0.7$ nm. The relaxation oscillation periods τ_{RO} in the laser are of the order of 100 ps, with the exact value of τ_{RO} depending on the injection current. The internal coupling delay $\tau = 5.7$ ps, which corresponds to the length of the passive section, is of the order of the photon lifetime.

The selected ITL exhibits rich dynamics. The three injection currents and the device temperature form a codimension 4 parameter space, wherein quite distinct dynamical regimes and bifurcations can be deliberately and reproducibly adressed. Three basic dynamical regimes are found in the ITL: stationary or cw emission, periodic selfpulsations, and chaotic emission. The left side of Fig. 2.4 shows typ-

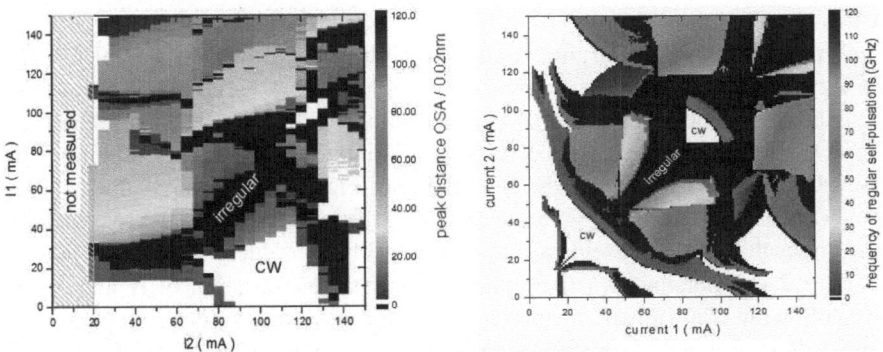

Figure 2.5: *Dynamical regimes of the ITL in the plane of the laser currents I_1, I_2. Left: measured, right: calculated. White: cw emission, gray: selfpulsations, black: chaos. $I_P = 0$, $T = 20\,°C$.*

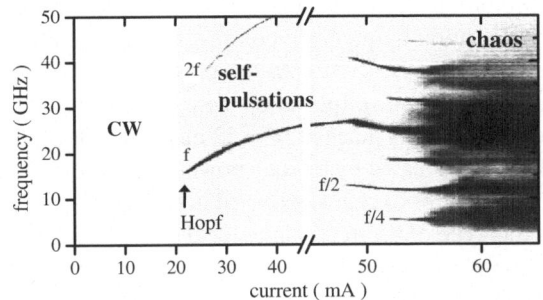

Figure 2.6: *Route to chaos in the freerunning ITL. Two period doubling bifurcations are followed by the occurence of chaos. Powerspectra of the emitted intensity are plotted in logarithmic grayscale versus I_P.*

ical optical spectra, i.e. Fourier transforms of the optical amplitude. However, the workhorse in the analysis of the dynamical regimes in the MSL are measured powerspectra, i.e. Fourier transforms of the emitted optical power (right side of Fig. 2.4), which show the (slower varying) frequency components in the radiofrequency range (in literature also referred to as 'rf spectrum'). Stationary emission is characterized by one dominant peak in the optical spectrum and a flat powerspectrum. Periodic selfpulsations are characterized by a comb of equidistant peaks in the optical spectrum, and by a sharp peak appearing at the pulsation frequency f in the powerspectrum. Usually, higher harmonics at multiples of f are also observed.

Chaotic emission is characterized by broad structures in the optical spectrum as well as in the powerspectrum. A cut through the 4 dim phase space of the ITL in the plane of the two laser currents is given in Fig. 2.5. The ITL shows large regions of cw operation (white), of selfpulsations with frequencies up to 30 GHz (gray) and of chaos (black). A corresponding simulated scenario considering identical DFB lasers is shown in the right panel of Fig. 2.5. Even though the lasers in the measured ITL device are not identical, the comparison of the two pictures illustrates that the measurements are indeed nicely reproduced by the simulations.

Fig. 2.5 is obtained for fixed phase current I_P on variation of I_1 and I_2. In the following control experiments, I_P is usually utilized as bifurcation parameter. When I_P is changed, the picture in Fig. 2.5 gets shifted along the bisecting line. A typical route to chaos observed upon variation of I_P is shown in Fig. 2.6. Measured power spectra of the laser output are plotted in grayscale along the vertical axis for different phase currents I_P. Starting with cw emission (flat spectrum), the laser first undergoes a Hopf bifurcation at $I_P = 22$ mA. Here, relaxation oscillations are undamped and a stable selfpulsation of $f = 16$ GHz frequency evolves. Going further, the pulsation frequency increases. A second harmonic at the double frequency appears (2f). At $I_P = 48$ mA, the pulsation loses stability in a period doubling bifurcation. Now, a selfpulsation with half frequency (f/2) takes over. Another period doubling takes place at $I_P = 52$ mA (f/4). Beyond $I_P = 56$ mA, the laser becomes chaotic, indicated by a broad powerspectrum. Such period doubling cascades are the main routes to chaos observed in this specific device. Unstable orbits generated in this type of bifurcation are ideal candidates for DFC [JBO+97a]. Therefore, above route to chaos will be the range of operation for the control experiments in Chapters 3 to 6. While virtually moving towards chaos in above picture, the control setup is applied to situations of increasing complexity.

Active Feedback Laser (AFL)

The second device utilized in the experiments is an AFL [BBK+02] (Fig. 2.7). This device combines a single DFB laser with ultrashort optical feedback. The laser section is a 1537 nm emitting single-mode index-guided DFB laser with a length of 200 μm. The two-part feedback cavity comprises a 550 μm passive waveguide and a 250 μm amplifying section (Fig 2.7). The laser facet is antireflection coated with a remaining power reflectivity of $R < 10^{-4}$, while the cleaved amplifier facet has a reflectivity of $R = 0.3$. The reflections at the interfaces are negligible. The roundtrip time in the feedback cavity is about 20 ps. The amplifying section enables a compensation for the optical losses which occur in a standard external cavity laser. This way, a higher - and tunable - feedback strength is achieved. Like in the ITL, the three sections are independently current biased, and the overall chip temperature can be controlled. The AFL also yields a codimension 4 parameter space. The divided feedback section allows for a more or less independent tuning of the internal feedback phase φ_P and the internal feedback strength K_i. K_i can be varied by tuning the current I_A in the amplifier section. Thereby, also the phase φ_P changes. The coupling phase φ_P in turn can be independently adjusted by a tuning of the injection current I_P on the passive section.

The dynamics of the AFL are described e.g. in [UWH$^+$05, BBK$^+$04, Bau04, Bro05] and [LGWH09]. Fig. 2.8 gives an overview of the dynamics in the plane of the amplifier current I_A and the phase current I_P. The AFL shows mainly cw emission (white) and selfpulsations with frequencies up to 40 GHz and beyond (grayscale). The chaotic regions are rather small. The AFL is studied additionally in this thesis (Chapters 3 and 7) because it undergoes both super- and subcritical Hopf bifurcations in a large parameter range.

2.2 Experimental Control Setup

Principles

The MSL devices introduced above shall be controlled optically by feedback from an external Fabry-Perot (FP) interferometer. This means that the laser output passes a FP interferometer and undergoes multiple roundtrips in the FP cavity. A part of the light is transmitted by the FP, another part is reflected back and reinjected into the laser as control signal. As already mentioned, optical phase shifts play a prominent role in this setup. This becomes clear when evaluating the expression for the optical signal reflected by a plane FP cavity. Actually, the Fabry-Perot interferometer is basic textbook knowledge. However, almost all books consider only the transmitted signal and omit the reflected part. Therefore, in Appendix B the optical amplitude (Eq. (2.1)) reflected by a plane FP cavity is derived.

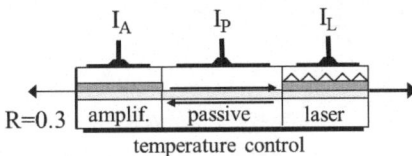

Figure 2.7: *Structure of active feedback laser (AFL), longitudinal cut.*

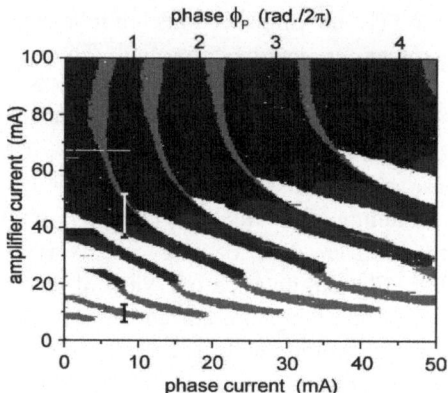

Figure 2.8: *Dynamical regimes in the active feedback laser. Areas of selfpulsations in the plane of phase and amplifier current, taken from [BBK⁺04]. DFB current is 70 mA. White: cw output. Grayscale: nonstationary output, the frequency components with highest intensity are coded as follows. Light gray: f < 15 GHz, dark gray: 15 GHz < f < 30 GHz, black: f > 30 GHz.*

In general, each component of the electromagnetic field emitted by a laser varies as $\mathscr{E}(t) = Re\left\{E(t)\, e^{-i\omega_0 t}\right\}$, with a fast optical carrier wave $e^{-i\omega_0 t}$ and a much slower varying complex amplitude $E(t)$ (Fig. 1.4 (b)). The dynamical scenarios in the MSL devices are dynamics of the slow amplitude $E(t)$. A periodic orbit corresponds to a periodic selfpulsation of the emitted power $P(t) = |E(t)|^2$ (Fig. 1.4 (b)). In this specific case, $\mathscr{E}(t)$ is a torus with two frequencies: the central optical frequency ω_0 in the femtosecond range and the much slower frequency of the optical power in the picosecond range. However, the two frequencies are not coupled, their timescales differ by orders of magnitude, and thus $\mathscr{E}(t)$ can be considered as modulated wave [Ran82]. Here, only the slow optical amplitude E is considered.

The reflected amplitude $E_b(t)$ in optical DFC is (Appendix B)

$$E_b(t) = Ke^{i\varphi} \sum_{n=0}^{\infty} (\mathscr{R}e^{i\phi})^n \left[\mathscr{F}e^{i\phi}E(t_{n+1}) - E(t_n)\right] \qquad (2.1)$$

with

$$t_n = t - \tau_l - n\tau.$$

Above expression resembles Eq. (1.2) for the control signal in extended delayed feedback control. The main control parameters in DFC reappear: the delay time τ, the feedback strength K and the memory parameter, here represented by $\mathscr{R}$, the effective reflectivity. $\mathscr{R} = Re^{-\alpha L}$ contains the mirror power reflectivity R and an additional term considering the cavity losses. The feedback strength K in Eq. (2.1) comprises a factor $\sqrt{R}$ corresponding to the reflection at the front mirror, and all propagation losses.

Eq. (B.9) contains three new parameters: the control loop latency time τ_l, which corresponds to the travelling time between laser and FP, and two optical phase shifts ϕ and φ.

A too large latency time τ_l can overrule the delay time τ, and reduces the control efficiency (see Appendix A). In the design of the experiment, the first goal is to achieve a minimal latency τ_l, which is of the order of τ, or below.

ϕ is the phase shift accumulated in the FP round trip. This quantity is governed by the phase velocity, while the propagation of the amplitude, and thus τ, are determined by the group velocity v_g of the FP medium. The value of ϕ determines the sign inside the brackets. A vanishing control signal (2.1) requires a minus sign. Thus, a careful adjustment and stabilization of ϕ is essential for noninvasive control.

The latency phase shift φ is accumulated in the latency roundtrip between laser and FP cavity. It enters the overall control gain $Ke^{i\varphi}$, turning it into a complex quantity. A nonvanishing latency phase φ causes a nondiagonal coupling of the control force (2.1) [RP04]. φ turns out as most crucial parameter in optical DFC. Thus, the second goal is to achieve a sufficiently controllable and stable latency gap. The impact of the quantities τ_l, ϕ and φ on expression (2.1) causes the main experimental challenges in the realization of optical DFC. In the following, the experimental conditions will be discussed in detail. Fig. 2.9 sketches the experimental control setup with the main control parameters.

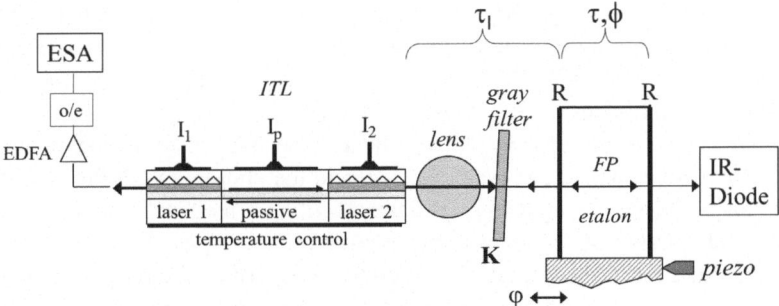

Figure 2.9: *Experimental setup for optical DFC.*

Operation of the MSL Chips

The MSL devices normally come as plug-and-play modules in a robust housing with optical and electrical connectors. In all-optical DFC, a major challenge is the reduction of the latency time τ_l. Thus, the FP must be placed extremely close to the MSL. For this reason, it is not possible to operate the lasers as modules here. Instead, direct access to one laser facet is necessary, requiring the MSL to be operated as bare chips (like in Fig. 2.2). At first, an environment has to be realized which serves for the operation of bare MSL chips and for the extraction of the optical output. This is quite fiddly, considering that the MSL chips are very small. Their overall length is 1 mm.

Current Supply and Temperature Control   The laser chips are mounted on copper heat sinks by the colleagues at the Heinrich Hertz Institute Berlin. The heat sinks have a width of 2 mm and a length of 6 mm and enable current injection to the different laser sections via integrated contacts in form of thin stripes. These contacts are adressed by specific conducting needles, which are separately adjustable by three PH100 needle holders (produced by SUSS MicroTec). A microscope supports a controlled manipulation of the current needles.

The temperature of the laser chip is controlled by an AD590 temperature sensor mounted inside the heat sink and a peltier element under the heat sink. The injection currents, temperature sensor and peltier element are controlled by a PRO 8000 laser diode control system. By these means, a setting of the injection currents and of the temperature with 0.01 mA and 0.01 K accuracy, respectively, is possible.

Light Extraction   The MSL device emits light on both facets. The emission has to be extracted for two purposes: for analysis, and in order to realize the feedback from an external FP cavity. Unfortunately, the emission shows large aperture angles of 20° and 40° FWHM in vertical and lateral direction, respectively. This requires thoroughly designed collimating devices.

At one facet, the emission is collected by an especially customized tapered glass fiber with a cone-shaped end (see Fig. 2.10 and the photograph in Fig. 2.2). Then it is transported via a single stage isolator, yielding a feedback suppression of $>$ 50 dB, and via fiber optic cables to the measurement devices.

At the other facet, the output is coupled to the external FP cavity. Minimization of the latency time requires a reduction of the dimensions of the collimating device as far as possible. Luckily, I was assisted by Dr. Güther from the Ferdinand-Braun-Institut für Höchstfrequenztechnik. Following his advice, a spherical sapphire ball lens with a diameter of 1 mm was chosen. Its optimal distance from the facet

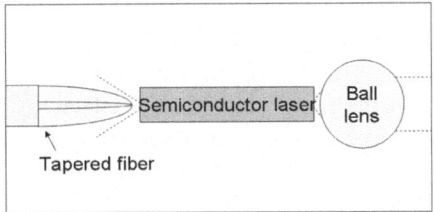

Figure 2.10: *Light extraction from the MSL chip. On the left side, the highly divergent output is coupled into a tapered fiber and used for analysis. On the right side, collimation with a sapphire ball lens enables the additional coupling of a FP cavity.*

was calculated by Dr. Güther to $(78 \pm 7)\,\mu$m. Without considering additional attenuating devices for the time being, this results in a lower border for the latency gap between laser and FP of $l = 1.83$ mm (in air), corresponding to a minimal latency time of $\tau_l = 1.2$ ps.

Fabry-Perot Etalons

Minimization of the latency gap also affects the type of FP interferometer which can be used. Solid FP glass etalons offer direct access to the FP front mirror and can thus be brought much closer to the laser facet than a FP interferometer consisting of two independent mirrors. In the latter case, the unavoidable mounting equipment of the mirrors considerably increases the latency. With solid etalons, the minimal interspace between laser and cavity is only limited by the necessary optics in the latency path. Another advantage of solid etalons is that they offer a stable FP roundtrip, so that no further attempts for stabilization of the FP phase ϕ are required.

The drawbacks of FP glass etalons are a finite absorption inside the cavity, a fixed roundtrip time τ and fixed FP resonance frequencies - the optical frequencies which are transmitted by the interferometer. To overcome these limitations, the FP etalons feature negligible absorption, and come in sets of several etalons with different τ. The quartz glass bears a deep absorption minimum of about $\alpha = 10^{-5}$ cm^{-1} (according to the producer) for the emission wavelength of 1550 nm. The lengths of the used etalons are adapted to the timescales of the MSL dynamics. They vary between $L_r = 3.5$ mm and $L_r = 15$ mm, corresponding to roundtrip times ranging between $\tau = 34$ ps and $\tau = 143$ ps, roundtrip frequencies between $f = 29.8$ GHz and $f = 7$ GHz, and free spectral ranges between $\Delta\lambda = 0.24$ nm and $\Delta\lambda = 0.06$ nm, respectively.

The mirror reflectivity R of the etalons is defined by an additional dielectric coating. R controls the number of roundtrips inside the cavity and thus the impact of states from further in the past. A too large reflectivity would result in very narrow reflectivity minima of the FP and can cause feedback-induced disturbances. Thus, moderate values $R = 0.3, 0.5$, and 0.76 are chosen. This yields a finesse $F \approx \pi R^{1/2}/(1-R)$ of $F = 2.46$, 4.44, and 11.41, respectively. Fig. 2.11 and

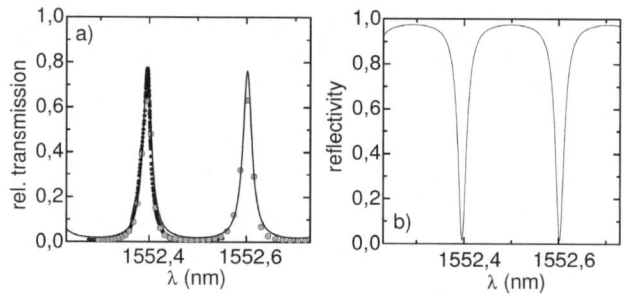

Figure 2.11: *Relative transmission spectrum (a) and calculated reflectivity spectrum (b) of a 4mm FP etalon with $R = 0.76$ mirror power reflectivity. (a): Dots: measured data, line: fitted spectrum.*

Fig. 3.2[1] plot measured transmission spectra of a 4 mm etalon with $R = 0.76$. The other FP parameters are determined by fitting an expression for the FP transmission to the experimental data (see Appendix B). The group refractive index is $n_g = 1.43$. The cavity losses are estimated to an upper border for the absorption coefficient of $\alpha = 0.05$ cm^{-1} and a mirror absorbance $A = 0.005$. The reflectivity of the mirrors is $R = 0.76$. This yields $\mathscr{F} = 0.98$, which corresponds to a remaining reflectivity of $4 \cdot 10^{-4}$ at the FP reflectivity minima.

Achieving Resonance between Laser Emission and Fabry-Perot

The essence of DFC is the noninvasivity of this control method. In few words, this means that the laser is subject to direct optical feedback, but the feedback power tends to zero. At first glance, such 'noninvasive optical feedback' possibly seems contradictory. As a matter of fact, conventional optical feedback from a simple mirror is always invasive and shifts the point of operation of the laser [LK80].

---

[1]The different levels of the transmission maxima in the two figures are most likely caused by different incidence angles of the beam.

However, direct feedback from a FP cavity can be adjusted to be noninvasive, featuring a vanishing reflected signal [TWWR06].

This fact follows directly from Eq. (2.1): $E_b(t)$ vanishes when the difference in the brackets becomes zero. This requires several conditions to be met at once. Assuming that the laser output is selfpulsating, i.e. a modulated wave with a T-periodic slow amplitude $E(t) = E(t+T)$, the DFC delay condition is

$$\tau = mT, \; m \in \mathbb{N}. \tag{2.2}$$

In optical DFC, the condition $\mathscr{F} e^{i\phi} = 1$ has to be fulfilled additionally. This requires that the FP phase shift ϕ meets the resonance condition

$$e^{i\phi} = 1, \tag{2.3}$$

and that the losses in the FP tend to zero:

$$\mathscr{F} \to 1. \tag{2.4}$$

In a real experiment, a FP without losses cannot be achieved. In principle, $\mathscr{F} = 1$ could be reached by a slight amplification inside the FP cavity, but this would involve an additional nonlinear element and is therefore not suitable here. Noninvasive control in the experiment can be defined weaker: the cavity losses should be minimized such that the effectively reflected amplitude $E_b(t)$ is below a level where the laser state is affected by the feedback. The value of $\mathscr{F}$ is determined by the selected etalon and cannot be influenced in measurement. However, conditions (2.2) and (2.3) are well controllable by the experimentalist.

The adjustment of T/τ and ϕ in order to meet conditions (2.2) and (2.3) can be illustrated in Fourier space. The optical intensity spectrum $|E(\omega)|^2$ of a selfpulsation with period T is a comb of equidistant lines, separated by a frequency shift $2\pi/T$ (dashed lines in Fig. 2.12). The reflectivity spectrum $|E_r(\omega)|^2/|E_i(\omega)|^2$ of a FP interferometer with internal roundtrip time τ has equidistant minima ('resonances'), separated by $2\pi/\tau$ (gray line in Fig. 2.12)[2]. The signal reflected by the FP vanishes at resonance when all frequency components of the optical emission fit to minima of the reflectivity spectrum. In case of an ideal interferometer, the reflected power is only limited by noise. If condition (2.2) is fulfilled, the spacing of the optical modes is an integer multiple of the spacing of the FP resonances, and thus a resonant adjustment is possible. The relative position of the comb to the FP resonances is defined via ϕ. At resonance, ϕ equals zero, and condition (2.3)

---

[2]The value of $\mathscr{F}$ steers the depth of the minima in the FP reflectivity spectrum. For $\mathscr{F} = 1$ they go to zero.

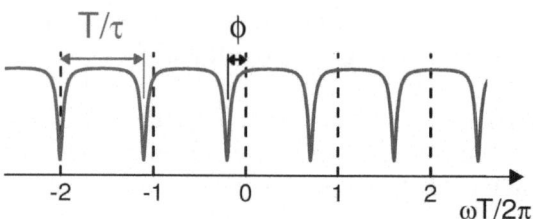

Figure 2.12: *Adjustment of a selfpulsating optical signal to a Fabry-Perot cavity. Optical spectrum of the selfpulsation (dashed lines) and Fabry-Perot reflectivity spectrum (gray line).*

is satisfied[3]. Noninvasive control requires Eq. (2.3) to be permanently fulfilled, making high demands on the stability of the FP phase ϕ.

Under resonant adjustment the reflectivity of the Fabry-Perot cavity is zero for the optical frequency components of the target state, but for all other frequencies a finite feedback is generated. This feedback to the off-resonant components causes the stabilizing effect in all-optical DFC, as it continuously drives the laser back to the desired state, back to minimal feedback.

The use of solid etalons prevents the adjustment of conditions (2.2) and (2.3) via the FP parameters. Fortunately, the MSL dynamics allow for a convenient adjustment via the laser state. The dynamical regimes are usually robust in a sufficient parameter range. Thus, the slow frequency of the amplitude as well as the optical wavelength are variable over a sufficient range without leaving the respective dynamical regime. Therefore, all adaptations can be realized by varying the point of operation in the laser. In the studied MSL, a temperature variation allows for a tuning of the emission frequency ω_0 - and thereby of ϕ - without affecting other properties like pulsation frequency or pulsation damping too much. Thus, equation (2.3) is usually fulfilled by adaptation of the temperature. The pulsation period T is always adjusted via a laser current. In all control experiments, the tracing of a stabilized state is performed along a line in parameter space with constant ω_0 and $1/T$.

In order to check for successful resonant adjustment, the control power $|E_b|^2$ reflected by the FP has to be monitored. A vanishing of $|E_b|^2$ indicates resonance. Unfortunately, a direct measurement is not convenient because stray light impedes

---

[3]The adjustment of a stationary state is simplified. Here, the optical spectrum contains only one frequency component, and the round-trip time τ is freely selectable. Only the relative position of the optical mode to the resonances must be defined via ϕ.

a satisfying identification of the zero point of $|E_b|^2$. Instead, the transmitted signal is monitored, and the laser emission is adjusted by a maximization of the transmitted power. A large-area infrared (IR) photodiode placed behind the FP detects the mean transmitted power $|E_t|^2$. From the level of $|E_t|^2$ it is possible to conclude for the power $|E_b|^2$ coupled back to the laser. Actually, only a 'relative FP transmission' can be measured. The detected 'relative FP transmission' is the spectrally integrated power transmitted by the FP, divided by the spectrally integrated power of the freerunning laser without FP, but in the same point of operation.

Variation of Control Parameters

Convenient values for the main control parameters in optical DFC - τ, R, K, and φ - are not known during the design of the experiment, but have to be determined by the experiments themselves. Thus, sufficient variability of these quantities is essential. The control delay τ and the memory parameter R can be varied only stepwise by the choice of the etalon. τ is variable between 33 ps and 140 ps. Three values for the reflectivity R are available: $R = 0.3, 0.5, 0.76$. The feedback strength K and the latency phase φ should be continuously variable during measurements.

Feedback Strength K   All-optical DFC only works in a finite range of K. The strength of the feedback must exceed a certain level in order to sufficiently influence the laser dynamics, but a too high K can lead to feedback induced disturbances. In the realized setup, K can be tuned over almost two orders of magnitude by a grayfilter with a density gradient. Optionally, an additional neutral density filter increases the range for K to almost four orders of magnitude.

The absolute value of K is determined by several parameters: the coupling efficiency of the lens, the damping introduced by the filter, and propagation losses: $K = K_{filter} \cdot K_{lens} \cdot K_{loss}$. Only K_{filter} can be determined experimentally. Thus, the exact value of K is uncertain. Instead, the different feedback strengths K_i used in different control experiments are compared with each other. This is realized by measuring the variation of the emitted mean power versus the latency phase φ in a fixed cw reference state[4]. In the experiments of Chapter 7, the feedback strength K is defined for once by elaborate comparison of measured and simulated data. In general, in most experiments K is of the order of $K \approx 0.05$.

---

[4] A more convenient determination of K by measuring the reduction of the laser threshold current was prevented here by the minimal stepsize of the current controller ($\Delta I = 0.01\ mA$), in conjunction with the low ITL threshold currents of about 10 mA. The resolution was not sufficient for a determination of the threshold reduction with satisfying precision.

Latency Time τ_l The thickness of the variable gray filter of 1.6 mm, and the fact that the filter must be slightly tilted to avoid feedback, increase the minimal latency gap to $l_{min} \approx 6$ mm, and the minimal latency time to $\tau_{l,min} \approx 40$ps. This value is of the order of the delay times τ in the setup. Therefore, the first condition on this setup is still satisfied.

Simulations show that not the complete range of τ_l is usable for control. Instead, the regions of successful noninvasive control versus τ_l are located around integer multiples of τ, and decrease in size with growing τ (see Fig. 4.10). The smallest realizable value is $\tau_l = \tau$. Thus, in all following experiments (except for Chapter 4) the latency time τ_l is always set equal to the delay time τ.

Latency Phase φ The phase shift φ accumulated in the latency round trip steers the algebraic sign of the reflected control signal (see Eq. (2.1)). Upon tuning of φ, the sign of $E_b(t)$ changes periodically between negative and positive. Thereby, the impact of the feedback changes between stabilization and destabilization. Control is possible in maximal half a phase period, thus φ must be controllable with high resolution. This requires a stabilization of the gap between laser and FP on a subwavelength scale.

φ is adjusted by a piezotranslation stage with a mechanical resolution of approximately 30 nm. For the experiments in Chapter 7, a position-stabilized piezotranslation stage with a resolution of less than 20 nm is used. Sufficient stabilization of the gap between laser and FP is achieved with an internally damped breadboard lying on a base frame of piled styrofoam with lightly inflated rubber hoses on top (thanks to advice from the staff of the coherence optics lab in the Department of Physics, Humboldt-University of Berlin). The damping capability of this construction was tested with a Michelson interferometer, and found sufficient. Fig. 2.13 shows a photograph of the realized experiment.

Analysis

The extracted laser emission is analyzed by an optical power meter, a FP spectrometer, and a 40 GHz electrical spectrum analyzer (ESA). Additionally, the power $|E_t|^2$ transmitted by the FP is measured. $|E_t|^2$ is detected by a large IR diode (2 mm by 2 mm area) placed behind the FP etalon.

Optical Power Meter   The power of the extracted optical signal is determined by a Noyes Fiber Systems OPM 5 optical power meter. The power meter detects the time-averaged absolute optical power: $\langle P(t) \rangle = \langle |E(t)|^2 \rangle$, with $E(t)$ the complex amplitude of the emitted optical field.

Electrical Spectrum Analyzer   The fast variations of the emitted power are measured with an electrical spectrum analyzer, which covers a frequency range of 9 kHz to 40 GHz. The optical output is first amplified by an erbium doped fiber amplifier (EDFA), and then converted to the electric domain by an ultrafast 50 GHz photodiode. The diode yields an electric signal $I_{PH}(t)$ proportional to the optical power: $I_{PH}(t) \sim \langle P(t) \rangle = \langle |E(t)|^2 \rangle$. This quantity is also time-averaged, but with an extremely short integration time - compared to a standard optical power meter - of less than 0.02 ns. The electrical spectrum analyzer measures the mean spectral

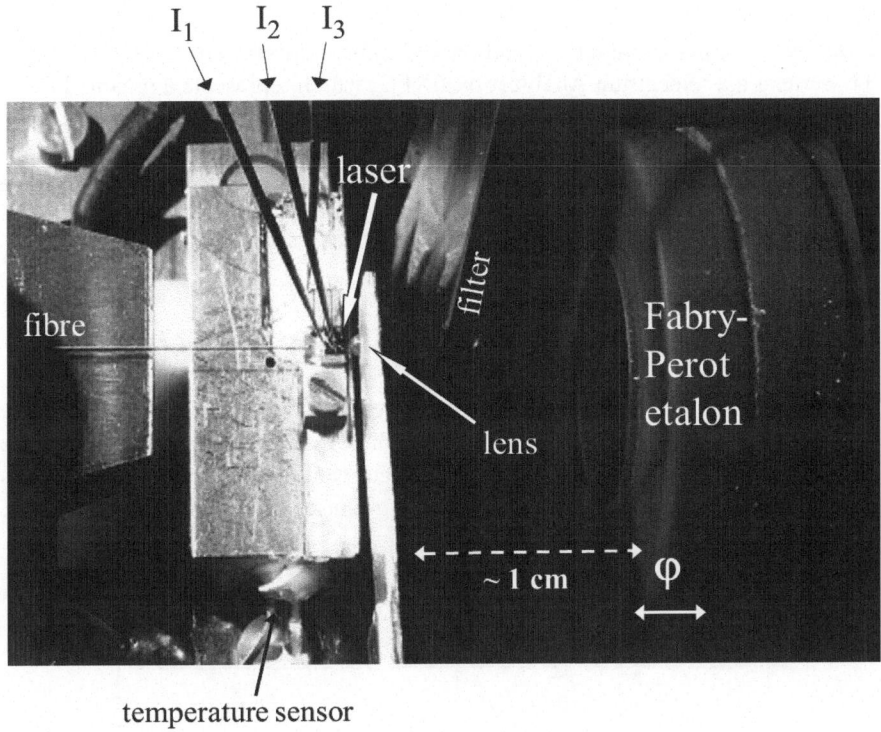

Figure 2.13: *Photograph of the experimental control setup. The bare laser chip is biased by three current needles ($I_1 - I_3$). The optical output at the left facet is coupled into a fiber and analyzed. At the right facet, the FP is coupled.*

density during the sampling time interval T (with t_0 the center of the interval):

$$I_T(\omega, t_0) = 1/T \int_{t_0-T/2}^{t_0+T/2} I(t) \cdot e^{-i\omega t} \, dt$$

of the converted electric signal in the radiofrequency (rf) range. In literature, this is referred to as 'rf spectrum', or 'powerspectrum'. The powerspectrum visualizes the frequency components of the emitted optical power in the detected range of 9 kHz to 40 GHz. Fig. 2.4 shows exemplaric powerspectra of the ITL dynamics. In this work, the dynamical regimes of the MSL are mainly analyzed via measured powerspectra.

Fabry-Perot Spectrometer   Additionally, optical spectra are measured with a TL Series Laser Spectrum Analyzer by EXFO, which represents a tunable Fabry-Perot spectrometer. Here, the square $|E(\omega)|^2$ of the fourier transform of $E(t)$ is determined. The optical output is sent through the FP spectrometer, while the frequency of the FP transmission maximum is tuned over a certain range. A diode behind the spectrometer detects the transmitted power in dependence on the frequency, thus generating an optical intensity spectrum.

2.3 Numerical Modelling of the Control Setup

The experiments are supplemented by numerical simulations of the control setup. This is worthwile as a relief and support for the experiment. The simulations are helpful in the prearrangement, where convenient values for the control parameters have to be determined. In the postprocessing phase, simulations enable a more

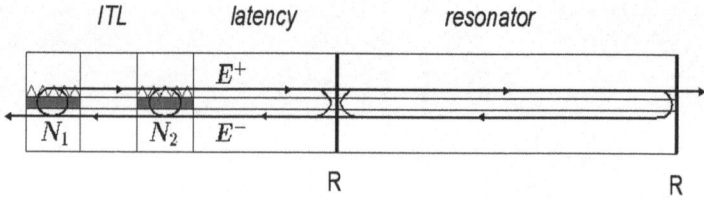

Figure 2.14: *Control setup in the numerical simulations.* N_1, N_2: *inversion in section 1 and 2,* E^+, E^-: *for- and backwards traveling optical amplitudes, R: reflectivity of the FP mirrors.*

comprehensive study of e.g. control domains. These purposes call for a very good reproduction of the real processes by the calculations. Luckily, I had access to a strong simulation tool. The dynamics of the used multisection lasers are excellently described in a traveling-wave (TW) model [Fle68, ZSFYMC94, HMM96, SM06, Rad06] in the framework of the Maxwell-Bloch equations [AB65].

The TW model can be easily extended to an additional description of the external FP feedback. The multisection laser is added by two passive sections which represent the latency gap and the FP cavity. Fig. 2.14 plots a sketch of the simulation setup with the controlled ITL. The FP is realized by defining a finite reflectivity R at the interfaces of the resonator section.

The optical field along the multisection cavity consists of superposed forward and backward traveling waves:

$$\mathcal{E}(z,t) \sim Re\left\{E^+(z,t)e^{-i(\overline{\omega}t-\overline{k}z)} + E^-(z,t)e^{-i(\overline{\omega}t+\overline{k}z)}\right\}. \tag{2.5}$$

$\overline{\omega}$ and $\overline{k} = \pi/\Lambda$ are the central frequency and Bragg wavenumber of the DFB sections, respectively. $E^+(z,t)$ and $E^-(z,t)$ represent the forward and backward traveling optical amplitudes, respectively. The field-carrier dynamics follow partial differential equations for the optical amplitudes, and a rate equation for the gain:

$$\left[\frac{1}{c_g}\frac{\partial}{\partial t} \pm \frac{\partial}{\partial z}\right]\begin{pmatrix}E^+ \\ E^-\end{pmatrix} = i\begin{pmatrix}\beta & \kappa \\ \kappa & \beta\end{pmatrix}\begin{pmatrix}E^+ \\ E^-\end{pmatrix}, \tag{2.6}$$

$$\frac{dn_s}{dt} = \frac{I_s}{eV} - \frac{n_s}{\tau_n} - \frac{c_g g_s S_s}{1+\varepsilon S_s} \quad (s=1,2). \tag{2.7}$$

Here, c_g denotes the group velocity, s the index of the setup sections, I_s the current injected into section s, e the elementary charge, V the volume of the active zone, τ_n the carrier life time, and ε the coefficient of nonlinear gain saturation.

The active DFB sections $s=1,2$ are implemented by setting the complex DFB coupling constant κ and the carrier density n_s non-zero here, and zero in all else sections. β is the propagation parameter of the waveguide relative to $\overline{k}$ and follows

$$\beta = \delta_s + i\frac{\gamma_s}{2} + \frac{g_s}{2}\left(\alpha - \frac{i}{1+\varepsilon S_s}\right), \tag{2.8}$$

with α the linewidth enhancement factor, and δ_s and γ_s the constant background contributions from refraction and absorption, respectively, at $\overline{\omega}$. In the active sections, the unsaturated optical gain g_s is approximated by

$$g_s = g' \cdot (n_s - n^{tr}), \tag{2.9}$$

with the differential gain g' and the transparency concentration n^{tr}. In all other sections g_s is set to zero. S_s is the effective photon density in section s:

$$S_s = (c_g \hbar \overline{\omega} V)^{-1} \int_s dz (|E^+|^2 + |E^-|^2). \tag{2.10}$$

The medium polarization is adiabatically eliminated here. Further, a mean-field-type approximation is used, assuming that the occupation inversion in the DFB laser sections can be described by spatially averaged carrier densities $n_s(t)$ [SHW$^+$06].

At all interfaces, linear boundary conditions are assumed:

$$\begin{pmatrix} E^+(+,t) \\ E^-(-,t) \end{pmatrix} = \begin{pmatrix} t_+ & r_- \\ r_+ & t_- \end{pmatrix} \begin{pmatrix} E^+(-,t) \\ E^-(+,t) \end{pmatrix}. \tag{2.11}$$

Here, arguments and subscripts $\mp$ refer to incidence from left- or right hand side. The incoming waves at the front and end facet are set to zero: $E^+(-,t) = 0$, $E^-(+,t) = 0$. $(t_\pm)$ and $(r_\pm)$ are the transmission and reflection coefficients, respectively, at the interfaces. They follow the power-conservation conditions

$$|t_\pm|^2 + |r_\pm|^2 = 1 - A_\pm \qquad \text{and} \qquad t_+ r_-^* = -t_-^* r_+, \tag{2.12}$$

with $A_\pm$ the interface absorbance.

The finite propagation velocity in the latency section and in the resonator section indeed realizes a delayed feedback in this model. The main control parameters, feedback strength K and latency phase φ, are described by

$$K = \exp(-\gamma_l L_l) \qquad \text{and} \qquad \varphi = 2\delta_l L_l. \tag{2.13}$$

L_l is the length of the latency section l. γ_l and δ_l are the above introduced absorption coefficient and static wavenumber, respectively, in the latency section. The reflectivity R is defined by the reflection coefficients at the resonator section interfaces. The reflection coefficients at the other interfaces are set to zero.

Above equations are solved with the software LDSL-tool, which was developed by Mindaugas Radziunas (see http://www.wias-berlin.de/software/ldsl/ for further information). Table 2.3 gives exemplaric parameters used in the simulations in Chapter 4, with triple values denoting the ITL sections [DFB1, P, DFB2] or the AFL sections [A, P, DFB]. Deviating parameter values used in the simulations in Chapters 5, 6, 7 are given there.

Table 2.1: Simulation parameters

Parameter	Value
group velocity	$c_g = c/3.8$
complex Bragg coupling	$\kappa = [1,0,1] \cdot (250 + 6i) \text{ cm}^{-1}$
line-width enhancement factor	$\alpha = [-5,0,-5]$
internal optical losses	$\gamma = [25,20,25] \text{ cm}^{-1}$
static wave-number at transparency	$\delta = [27978, \text{variable}, 17910] \text{ m}^{-1}$
differential gain	$g' = [9,0,9] \cdot 10^{-21} \text{ m}^{-1}$
gain compression factor	$\varepsilon = [5,0,5] \cdot 10^{-24} \text{ m}^3$
transparency concentration	$n^{tr} = [1,0,1] \cdot 10^{24} \text{ m}^{-3}$
injection currents	$I = [8,0,80] \text{ mA}$
section lengths	$L = [220,440,220] \text{ } \mu\text{m}$
cross section of active zone	$\sigma = 4.5 \cdot 10^{-13} \text{ m}^2$
elementary charge	e
nonlinear recombination rates	$n/\tau_n = An + Bn^2 + Cn^3$
A	$A = 3 \cdot 10^8 \text{ s}^{-1}$
B	$B = 1 \cdot 10^{-16} \text{m}^3 \text{s}^{-1}$
C	$C = 1 \cdot 10^{-40} \text{ m}^6 \text{s}^{-1}$

2.4 Summary and Comparison with Literature

An experimental setup for all-optical DFC of multisection semiconductor laser
(MSL) chips is realized. The experimental environment enables the operation of
bare MSL chips, and the addition of direct feedback from an external plane Fabry-
Perot (FP) cavity to the emission of one laser facet. The FP interferometer is
realized as a solid etalon. This choice requires an indirect adjustment of the optical emission to the fixed FP, but ensures sufficient stability and a conveniently
short travelling time τ_l between laser and FP. This so-called latency time can be
reduced to $\tau_l = 40$ ps, which is of the order of the delay time τ. This is particularly
important, as a too large latency would decrease the ability for control. Further,
much effort is made to ensure control and stabilization of the latency gap on a
subwavelength scale. This is essential in optical DFC, where the phase shift φ accumulated in the latency gap presents a crucial control parameter. By means of this
phase-sensitive direct optical feedback from a plane FP cavity, different types of
unstable states embedded in the dynamics of the MSL devices shall be stabilized.

On the first glance, the realized setup could possibly be confused with other approaches already reported in literature. Stabilization of lasers by external cavities is well-established [WH91, NKC04]. FP cavities can be used e.g. as narrow spectral filters placed inside a laser resonator to achieve smaller linewidth and stable emission frequency. Another method exploiting external FP cavities bases on a filtered feedback from spectral components at the resonances [DHD87, LCB89, VIV03]. This approach seems to be quite similar to all-optical DFC. Here, a semiconductor laser is coupled optically to an external confocal off-axis operated Fabry-Perot cavity, and receives feedback only from the cavity resonances. Under proper adjustment, the laser self-locks to the resonance and emits single mode output with narrower linewidth and constant (locked) optical frequency. In contrast to all-optical DFC, the applied feedback from the confocal Fabry-Perot cavity features a power maximum at resonance, i.e. the scheme works highly invasive. Optical DFC works in a power minimum. Lately, semiconductor lasers with so-called filtered optical feedback received attention [FAY$^+$00, EKL$^+$06, ELK$^+$07, RMD$^+$99, FYL$^+$04, YK88]. Here, the laser receives feedback from the signal transmitted by a plane FP. This scheme also works highly invasive. Only in the powerful Pound-Drever-Hall technique [DHK$^+$83], the laser emission is locked to a FP reflectivity minimum. But here, the stabilization is realized via an electronic feedback loop.

3 Stable States with Resonant Fabry-Perot Feedback

The experimental setup is first used to control properties of stable states. In above sentence the term 'control' is a little bit misleading, as it refers to a context, which is different from the one used so far in this work. 'Control' in the present Chapter does not refer to the stabilization of an otherwise unstable state, but to the deliberate influencing of properties of stable states. From the experimental point of view, stable states present the simplest situation to start with. Here, the parameters of the state, which determine a resonant coupling of the FP - the central wavelength ω_0 and, optionally, the pulsation frequency T - are known. On the one hand, the functionality of the setup can be checked by experiments with stable states. Then again, the control of certain properties of stable states is interesting itself, especially in connection with noise. An important example is the control (improvement) of oscillator coherence. Controlling properties like the damping and frequency of relaxation oscillations (RO) in a laser is also of interest, because these quantities determine important laser parameters like small signal response, relative intensity noise, and chirp [AD93].

In what follows, the optical control of a stable stationary state (Sec. 3.1) and the control of coherence (Sec. 3.2) is studied.

3.1 Stationary States

A cw emitting laser subject to resonant feedback from a FP cavity is considered. The laser is set to stable stationary emission and the wavelength is adjusted to be resonant with a FP reflectivity minimum. Then, the response to parameter variation is observed.

This case has been studied theoretically in the framework of generalized Lang-Kobayashi equations in [TWWR06]. This extensive work, which has not been evaluated experimentally so far, yields several results which are important here. First, the resonant FP feedback does not change the stationary laser emission (called 'solitary laser mode'), but strongly affects its stability properties. The impact of the feedback changes qualitatively when exceeding a critical feedback

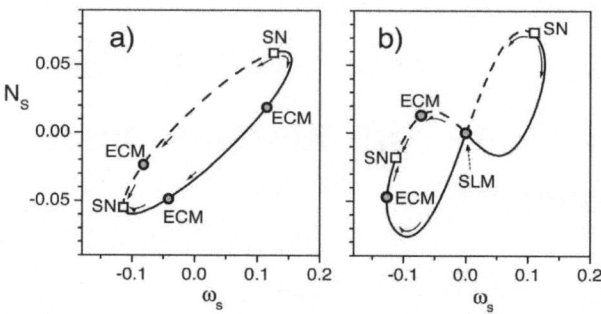

Figure 3.1: *Figure from [TWWR06]. External cavity modes (ECM) in the (ω, N) plane for conventional optical feedback (a) and feedback from a resonant FP cavity (b). The feedback strength is K = 0.05. SN: Saddle-node bifurcation, SLM: solitary laser mode. Solid line: node, dashed line: saddle. The external cavity mode move along these curves when the feedback phase changes. Arrows show the direction of movement with growing feedback phase. The solitary laser mode does not change its position with phase.*

strength K_{crit}. Below K_{crit}, only the completely unchanged cw emission of the laser exists. However, the stability of the solitary laser mode changes drastically in dependence on the feedback phase φ. On tuning φ, the small signal dynamics of the laser can be strongly influenced. When subject to a small disturbance, a cw emitting laser without external feedback shows damped relaxation oscillations (RO) back to the stationary state[1]. The effects of additional FP feedback range from weak damping of the RO to overdamping, and also undamping, resulting in a destabilization in a Hopf bifurcation. Above K_{crit}, additional feedback-induced stationary states appear (called 'external cavity modes'). The stability of the solitary laser mode is reduced in this regime by multistability and transcritical bifurcations. The external cavity modes live on a tilted figure eight in the plane of inversion N and optical frequency ω, with the unchanged solitary laser mode in the center (see Fig. 3.1 (b)). The images in Fig. 3.1 illustrate the fundamental difference between conventional optical feedback from a mirror and FP feedback. In the case of conventional optical feedback from a simple mirror, the external cavity modes live on a tilted ellipse in (N,ω) plane (Fig. 3.1 (a)). The unperturbed state in the center does not exist here due to the invasive nature of conventional mirror feedback. The study [TWWR06] proves that indeed 'noninvasive' optical feed-

---

[1]To be precise, only lasers operating in the 'class B regime' show relaxation oscillations. The lifetime of the upper state in the laser must exceed the cavity damping time.

back - in the sense that an unperturbed state exists - can be realized with a resonant
FP interferometer. Further, the crucial role of the feedback phase φ (called 'latency phase' in this thesis) is underlined. Maximum stability of the solitary laser
mode is found for values of φ different from π (which corresponds to standard
DFC).

Now I turn to the experimental verification of the configuration considered
above: the ITL in stable cw operation receives resonant feedback from a FP etalon.

Controllability of φ   At first, the question is adressed whether the latency phase
φ is sufficiently controllable in the all-optical setup. A FP etalon with $1/T = 26$
GHz and $R = 0.76$ is added to the stationary ITL. The optical emission from one
ITL facet is reflected by the FP and reinjected into the laser. For the time being, the
feedback strength K is set to the order of 10^{-5} to avoid any feedback to the laser.
The optical frequency is tuned via the laser temperature, and the power transmitted
by the FP is observed meanwhile. This yields the transmission spectrum shown
in Fig. 3.2. The emission wavelength is then adjusted equal to the transmission
maximum, and the feedback strength is increased to $K = 0.04$. Now, the laser
emission is adjusted resonant to the FP etalon.

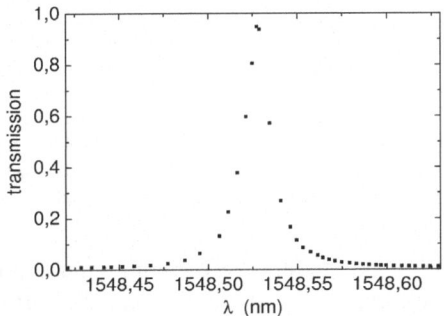

Figure 3.2: *Measured transmission spectrum of the 26 GHz etalon. The mirror reflectivity
is $R = 0.76$.*

Then, the FP is shifted towards the laser by increasing the piezo voltage in small
steps, corresponding to a decrease of φ. The laser remains in cw operation over the
whole range, but a periodic variation of the emitted power is observed. For each
measuring point, the power emitted by the laser at the opposite laser facet, and
the power transmitted by the FP are recorded. Fig. 3.3 plots the relative FP trans-

mission[2] (a) and the absolute emitted laser power $\langle |E(t)|^2 \rangle$ (b) versus the piezo voltage. The emitted laser power shows constant plateaus of maximal transmission and minimal emitted power, which cover a range of about 0.2 of one phase period. Inbetween, the power is increasing and the transmission falls.

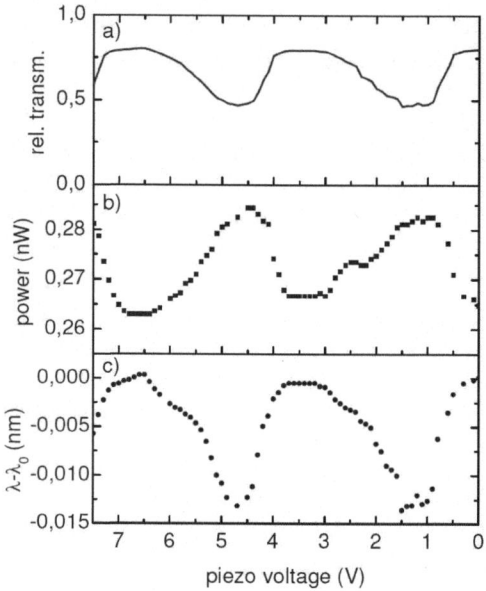

Figure 3.3: *Tuning of latency phase φ after the FP is coupled resonantly to the cw emitting laser. (a) Relative FP transmission and (b) optical power at the opposite laser facet versus the piezo voltage U_{piezo}. $(I_1, I_P, I_2) = (49, 0, 11)$ mA, $K = 0.04$.*

A similar behavior of the modulus of the emitted amplitude $|E|$ versus φ has been derived in [TWWR06]. Fig. 3.4 taken from [TWWR06] plots a bifurcation diagram versus φ for fixed K. $|E|$ has a constant plateau where the solitary laser mode ('SLM') is stable. In a transcritical bifurcation (black triangle at -0.8 in Fig. 3.4), the solitary laser mode loses stability in favor of a feedback-induced external cavity mode ('ECM'). In the external cavity mode region, the value of $|E|$ depends on φ, and is higher than for solitary laser mode operation.

Thus, the constant plateaus of the emitted power observed in Fig. 3.3 (b) correspond to regions of a stable solitary laser mode. This is supported by the maximal

---

[2]Please note that here, for once, the term 'relative FP transmission' refers to the power transmitted by the FP divided by the laser power at the opposite facet.

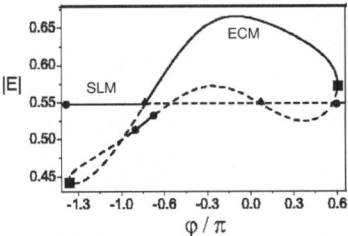

Figure 3.4: *Bifurcation diagram in the $(|E|, \varphi)$ plane for $K = 0.05$ (from [TWWR06]). Square: saddle-node bifurcation, triangle: transcritical bifurcation, circle: Hopf bifurcation. Solid line: stable stationary solution, dashed line: unstable solution.*

transmission observed simultaneously. In the other regions, external cavity modes are lasing. In [TWWR06], external cavity modes appear when K exceeds a critical value. For this setup, a feedback strength $K = 0.04$ is already beyond the critical value.

Summarizing, the experiment demonstrates that the latency phase φ can be sufficiently controlled, and that the laser with resonant FP apparently can show feedback-induced states. Only a part of the phase period can be used for noninvasive control. In above experiment, the solitary laser mode is lasing in a range of 0.2 of the phase period. For all following control experiments it is prerequisite that the latency phase is set to the usable range. A resonant adjustment of the FP etalon will always include the adaptation of φ, in addition to the adaptation of the optical frequency, and, optionally, of the pulsation frequency. The adaptation of φ is nontrivial, because φ depends on the optical carrier frequency: $\varphi = \omega_0 \tau_l$. When the carrier frequency ω_0 of the laser is tuned to a fixed FP resonance, the latency phase φ is inevitably detuned. Readjusting φ in turn affects the adjusted laser state, and so on. Whether φ is in the usable range will be checked for by monitoring the power transmitted by the FP, which has to be maximal, and by observing measured power spectra to identify the dynamical regime of the laser.

Controlling the Stability of a Stationary State   A central result in [TWWR06] is that resonant FP feedback strongly affects the stability of a stable stationary state. A measure for the stability of a stationary state is the damping of the RO. Close to bifurcations, the stationary state can get less stable and the RO damping decreases. Here, omnipresent random fluctuations can induce self-oscillatory motion [NSS97], which is indicated by a weak broad peak appearing in the power-spectrum at the RO frequency. This is the so-called noise-induced precursor of the

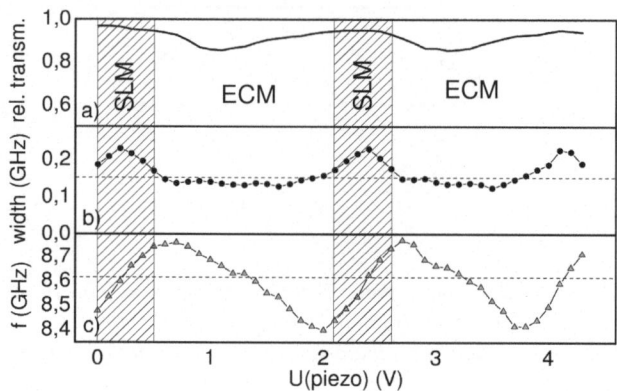

Figure 3.5: *Control of damped relaxation oscillations by resonant FP feedback. (a) relative FP transmission, (b) RO peak width, and (c) RO frequency, vs φ. Dotted lines: freerunning values.* $1/\tau = 17.5$ *GHz,* $R = 0.5$, $\tau_l \approx \tau$.

bifurcation (see e.g. gray line in Fig. 3.6 (a) and (b)). The width of this precursor is proportional to the RO damping, and, thus, a measure for the stability of the cw state.

The influence of FP feedback to such a regime is evaluated experimentally. The AFL is driven in cw operation close to a Hopf bifurcation. Here, a distinct precursor appears in powerspectrum (gray line in Fig. 3.6 (b)). A FP etalon is added to the AFL with a feedback strength $K = 0.03$. The laser emission is adjusted to resonance according to the description in the previous paragraph.

Tuning of φ now yields a periodic variation of the damping and of the frequency of the RO (Fig. 3.5). In the regions with high FP transmission the RO damping is increased (shaded regions in Fig. 3.5 (b)), indicating an improvement of the stability of the cw state. In the other regions, the transmission is lowered. This is the range of external cavity mode operation. The FP feedback also affects the frequency of the damped RO (Fig. 3.5 (c)), which varies periodically in a range of about 5% around the freerunning value.

Above experiment is an optical example for the control of noise-induced oscillations by DFC, which has been studied theoretically in the last years [JBS04, BJS04, PAS05]. This aspect gained interest, because the suppression of noise-induced oscillations in neural brain dynamics is an important clinical problem in the therapy of e.g. Parkinson's disease, essential tremor, and epilepsy [TRW+98, Tas02, GCB02]. The improvement of the stability of a stable stationary state within

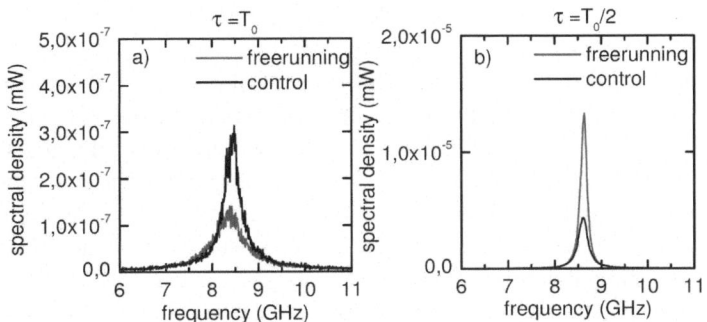

Figure 3.6: *Control of damped RO in dependence on the delay* τ. *(a) Enhancement of the oscillations with* τ = T_0. *AFL currents:* $I_L = 99.9$ *mA,* $I_P = 31.3$ *mA,* $I_A = 9$ *mA.* $1/\tau =$ *8.3 GHz. (b) Suppression of the oscillations with* τ ≈ T_0/2. *AFL currents:* $I_L = 99.9$ *mA,* $I_P = 32$ *mA,* $I_A = 9$ *mA.* $1/\tau = 17.5$ *GHz. Feedback strength K = 0.03. R = 0.5. FP phase and latency phase are adjusted to achieve minimal FP reflection. Gray: freerunning laser, black: with resonant FP etalon.*

the shaded regions in Fig. 3.5 represents a preliminary stage of a stabilization of an unstable stationary state, which will be studied in Chapter 4.

The Choice of the Delay Time τ   In the previous measurement, a FP delay time τ = 57 ns was chosen for controlling a stationary state which shows damped RO with a period $T_0 = 116$ ns. Which role does the choice of τ play here? When stabilizing a stationary state by DFC, the delay τ can be (almost) freely selected. The only critical choice is τ = T_0 [HS05]. Fig. 3.6 compares the choices τ = T_0 and τ = $T_0/2$. In both experiments, the stable cw emission is resonant to a FP reflectivity minimum, and φ is adjusted to achieve maximal transmission. For τ ≈ $T_0/2$, the RO are stronger damped (Fig. 3.6 (b)), and the stability of the stationary state is improved. The contrary choice τ = T_0 decreases the stability of the involved stationary state, and improves the regularity of the occurring RO (Fig. 3.6 (a)) [BJS04, FS07]. The latter aspect is referred to in literature as 'coherence control', and will be further considered in the following section.

3.2 Controlling the Coherence of Oscillations

The last experiment touched a field of control theory which has not been mentioned so far, and which is actually beyond the scope of this thesis: the control of coherence by DFC. Nevertheless, it yields interesting aspects also for the present work, and will thus be touched on briefly.

In the last years, the capability of DFC to control properties of oscillations has been studied. In this context, the term 'control' refers to the specific goal of an enhancement of the coherence, that is, the constancy of the frequency of oscillations [GRP03, BAM04, PP06a]. There are several reasons why coherence control is of significance. Oscillation coherence determines the quality of (optical) clocks and lasers. Further, increasing the coherence of a self-sustained oscillator also enlarges its ability to phase-synchronize with external signals [PRK01, GRP03, BAM04]. Phase-synchronization in turn is an omnipresent natural phenomenon, e.g. in spatially extended ecological systems [BHS99], with the heartbeat [SRKA98], or in the muscle activity of a Parkinsonian patient [TRW+98]. It can become a potential basis of chaos communication [VR98]. There is one specific aspect which makes coherence control interesting especially for the present thesis: it represents a preliminar stage of the control of an unstable periodic orbit [BAM04].

Control of Coherence: Phase Model   The previous section adressed the importance of the choice of τ when controlling the regularity of oscillations. This aspect is illustrated theoretically in [GRP03]. The coherence of an oscillator with frequency Ω_0 can be quantified by the phase diffusion constant $D_0 \propto \langle [\Phi(t) - \langle \Phi(t) \rangle]^2 \rangle / t$, with the uniformly growing phase $\Phi(t) \approx \Omega_0 t + \Phi_0$. Here, angle brackets refer to a suitable time average over the period of oscillations in order to keep only terms depending on phase differences [GRP03]. Φ_0 is the initial phase shift. The diffusion constant D_0 is proportional to the linewidth of the oscillation

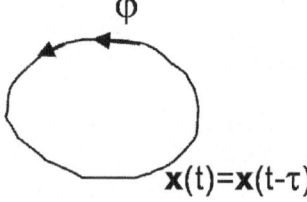

Figure 3.7: *Periodic orbit in phase space. Perturbations acting on the phase Φ influence the movement along the direction of the orbit.*

peak in the powerspectrum. Weak perturbations in first approximation only affect the phase of an oscillator, not the amplitudes. Thus, weak noise mainly affects the movement of a trajectory along an orbit (Fig. 3.7). In [GRP03], the authors derive an equation for the phase Φ of the noisy Van der Pol oscillator subject to feedback with a delay τ:

$$\dot{\Phi} = \Omega_0 + K/2 \cdot \sin[\Phi(t-\tau) - \Phi(t)] + \varepsilon(t), \qquad (3.1)$$

with Ω_0 the freerunning oscillation frequency and $\varepsilon(t)$ the effective noise. In a linear approximation assuming weak phase fluctuations and a weak control force, the diffusion constant D of the controlled oscillator is obtained as

$$D = D_0 \cdot (1 + K/2 \cdot \tau \cos(\Omega_0 \tau))^2. \qquad (3.2)$$

The impact of the feedback term obviously depends on the sign of $\cos(\Omega_0 \tau)$. This sign can be changed by variation of the delay time τ. A delay τ which is a multiple of the oscillation period leads to a decreased diffusion constant, and an improved coherence. When τ is an odd multiple of half the oscillation period, D is increased, and the coherence decreases. Thus, DFC can both suppress or enhance phase diffusion depending on the choice of τ.

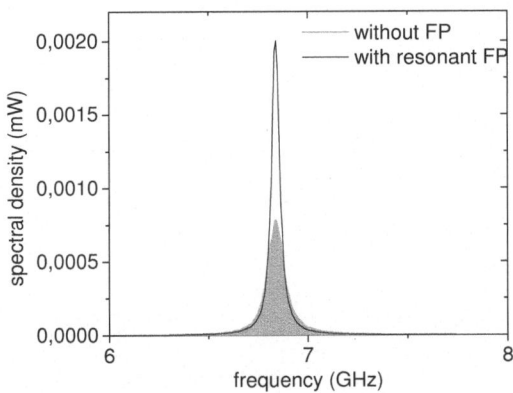

Figure 3.8: *Coherence improvement by resonant FP feedback. $(I_L, I_P, I_A) = (90.2, 18.1, 8.6)$ mA, $T = 19.4°C$. $K \approx 0.05$, $\tau_l \approx \tau$. Gray: powerspectrum of the freerunning selfpulsation, black: powerspectrum with resonant FP.*

All-Optical Control of Coherence   In the following, a stable selfpulsation (SP) with resonant FP feedback is considered. The AFL is operated in a regime of stable selfpulsations with about $1/T = 6.85$ GHz frequency. In the powerspectrum, a slightly broadened peak with a width of $\Delta\nu = 0.09$ GHz appears (gray filled in Fig. 3.8). A FP etalon with $1/\tau \approx 6.84$ GHz roundtrip frequency and $R =$ 0.5 mirror reflectivity is added to the laser with a feedback strength $K = 0.05$. The laser emission is carefully adjusted to resonance by iterative adaptation of T, ω_0, and φ. T is tuned via the amplifier current I_A, and the optical frequency ω_0 is adjusted via the laser temperature. At resonance, an improvement of the coherence is observed, accompanied by maximal FP transmission. The width of the oscillation peak in powerspectrum is decreased to $\Delta v_{FP} = 0.05$ GHz with FP feedback (Fig. 3.8).

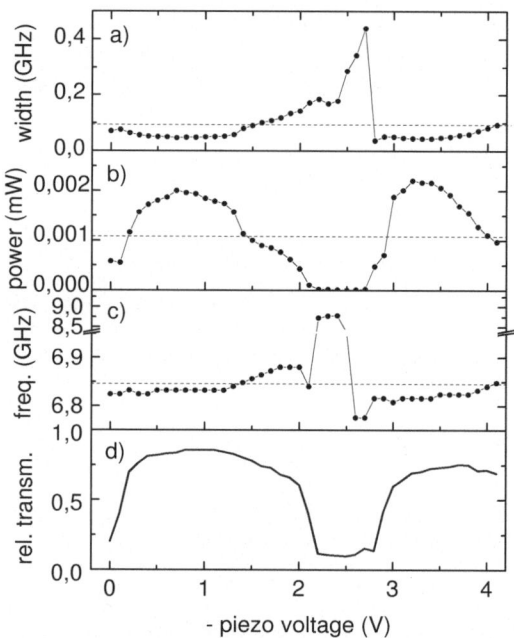

Figure 3.9: *Coherence control in dependence on the latency phase. RO peak width (a), height (b) and frequency (c), and the corresponding relative FP transmission (d) versus the piezo voltage. Dotted lines: freerunning values. Arrow: point of operation of Fig. 3.8.*

Then, the latency phase is varied by decreasing the piezo voltage in 0.1 V steps with otherwise fixed parameters. Fig. 3.9 summarizes the dependence of the main RO parameters on φ. The hatched area marks the regions of coherence control. Under control, the peak width is decreased compared to the freerunning value.

At the same time, the RO amplitude is increased and the relative transmission is high, indicating a minimal reflected signal. These regions cover 0.3 to 0.4 of a phase period. They are separated by a region of low FP transmission where the RO are suppressed by the feedback. Please note that under coherence control the RO frequency in Fig. 3.9 (c) is slightly below the freerunning frequency. This is in agreement with the findings in [GRP03].

Thus, the optical setup is in principle capable of controlling the coherence of pulsations. The effect is moderate for the used feedback strength of $K = 0.05$: the peak width is halved under control. A more comprehensive experimental investigation of all-optical coherence control was set aside here. Most probably, stronger effects of all-optical feedback on diffusion properties are possible for higher values of K, or for smaller values of τ [GRP03]. However, a systematic investigation was beyond the scope of this thesis.

3.3 Summary

The all-optical control setup is first tested in experiments with stable states. Resonant feedback from a plane FP cavity is applied to a multisection semiconductor laser. It is probed, if and how the FP feedback can influence properties of stable stationary states and of stable selfpulsating states.

First of all, these experiments prove the functioning of the control setup. Resonant feedback from a FP cavity has a clear impact on the laser dynamics, and especially a deliberate adjustment of the latency phase φ is possible. This is of utmost importance, because only a finite range of φ can be used for noninvasive control. In the remaining range of the phase period feedback-induced states appear.

Resonant feedback from the FP can improve or deteriorate the stability of a stationary state, depending on the choices for τ and φ. The two opposite choices of $\tau \approx T_0$ or $\tau \approx T_0/2$ lead to a decrease or increase, respectively, of the stability of the stationary state. Here, T_0 is the period of the damped relaxation oscillations (RO). A strong dependence of the stabilizing effect on the latency phase φ is observed. In a certain phase range, the damping of the noise-induced RO can be significantly increased by resonant FP feedback. This is the preliminary stage of the control of an unstable stationary state, which is considered in the next Chapter.

All-optical DFC is also capable of controlling coherence, i.e. the constancy of the frequency of oscillations. This is observed for noise-induced damped RO close to a bifurcation, as well as for stable selfpulsations. The oscillation coherence can be increased within a certain range of φ, characterized by a constant pulsation frequency and minimal FP feedback. This represents the preliminary stage of the control of unstable selfpulsations, which is studied in Chapters 5, 6, and 7.

4 Control of an Unstable Stationary State

So far, the improvement of the stability of an already stable stationary state by the all-optical setup is demonstrated. In a next step, an unstable stationary state subject to FP feedback is studied. In this experiment, the optical feedback changes the stability of the steady state, making it the first 'real' control experiment with the all-optical setup.

The stabilization of unstable stationary states, e.g. in lasers, is of immense practical interest. Often, a stable laser operation is required, and selfpulsations or chaos should be suppressed. Consequently, in literature a lot of approaches for the stabilization of unstable steady states are found, some are related to DFC. The method of 'occasional proportional feedback' is widely used to stabilize unstable steady states [RMM+92, JTH93], e.g. in a multimode Nd:YAG doubled laser [GIR+92, CS95, CRW94]. A deduced method called 'derivative control method' has been used to stabilize a multimode Nd doped optical fiber laser [BBDG93] and an electrochemical cell [PMR+99a]. Like DFC, it has the advantage of not requiring a reference state, and it is applicable to high-speed systems. But, the method suffers from high-frequency instabilities, and (extended) DFC has proven more robust in comparison [CBH+98]. Another method related to DFC is the 'washout filter', a highpass filter which can stabilize unstable steady states, e.g. in a CO2 laser [CLMG99]. The application of the original DFC scheme to unstable stationary states is studied in [NPT95, SG98, AP04, HS05, YWHS06, DHS07]. Experimental applications range from a Mackey-glass system [NPT95] to electrochemical systems [PPKH02], and a chaotic frequency doubled Nd-doped yttrium aluminium garnet laser [AP04]. However, all these experiments employ electronic feedback loops, and all-optical realizations are not reported.

The functional principle of DFC of an unstable stationary state can be illustrated by a linearized model that considers only the local properties of the target state [Ott]. The local stability properties of a stationary state $\vec{x}_*$ in an n dimensional system are governed by n characteristic multipliers with Floquet exponents $\Lambda_{0,k}$, which are either real, $\Lambda_{0,k} = \lambda_k$, or occur in pairs of complex conjugates, $\Lambda_{0,k} = \lambda_k \pm i\omega_k$. The Lyapunov exponent λ_k and the intrinsic eigenfrequency ω_k are real quantities reflecting the expansion from $\vec{x}_*$ and the revolution around $\vec{x}_*$, respectively. An unstable stationary state has at least one Lyapunov exponent $\lambda_k > 0$. The corresponding value of ω_k steers how a system trajectory departs from $\vec{x}_*$ in

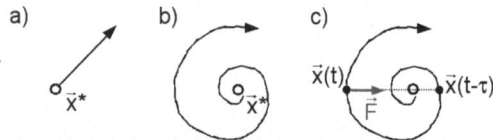

Figure 4.1: *Unstable stationary states in 2-dim phase space. a) Torsionfree state $\vec{x}_*$ with $\omega = 0$. b) Focus $\vec{x}_*$ with finite ω. c) DFC of unstable focus with delay τ chosen equal to half the intrinsic focus period. F: control force.*

this dimension. In case of a real positive $\Lambda_{0,k}$ with $\omega_k = 0$, solutions leave $\vec{x}_*$ straight in one direction (Fig. 4.1 (a)). For a complex conjugate pair $\Lambda_{0,k} = \lambda_k \pm i\omega_k$ the solution would spiral in a plane around $\vec{x}_*$ (Fig. 4.1 (b)). In the latter case, $\vec{x}_*$ is called a 'focus'.

Fig. 4.1 (c) demonstrates the functional principle of DFC of an unstable focus. One question is how to choose the delay time τ when controlling a stationary state. The dynamic system departs from an unstable focus along spirals, with the intrinsic focus period T_0. The control force F should suppress the outwards spiralling movement. The optimal choice for the delay time τ_{opt} is equal to half the intrinsic focus period T_0 [HS05],

$$\tau_{opt} = T_0/2 = \pi/\omega. \tag{4.1}$$

With this choice, a control force $\vec{F} = K(\vec{x}(t - \tau) - \vec{x}(t))$ (gray arrow in Fig. 4.1 (c)) always points towards the origin of the spiral, to the unstable focus. The outwards spiralling movement of the solution is suppressed.

All-optical DFC of a stationary laser state has been suggested early (e.g. in [Gau98]) and was extensively analyzed with a generalized Lang-Kobayashi model in [TWWR06] (see also Sec. 3.1). The all-optical control of an unstable focus presents the first step towards optical chaos control. It enables a control experiment under noncritical conditions. When operating close to the destabilizing bifurcation, the wavelength of the unstable focus and its intrinsic period T_0 are still known approximately, and the focus is only weakly unstable.

I realized the all-optical control of an unstable focus successfully in experiment, and identified the optical phase as new crucial control parameter compared to standard DFC. The results were published in [SHW$^+$06] and [WSH08]. In this Chapter, the experimental findings are summarized (Sec. 4.1), and a more comprehensive numerical investigation of the control parameter space is given (Sec. 4.2).

4.1 Stabilization of an Unstable Focus in Experiment

The Freerunning Bifurcation

The ITL is set to a regime shortly after a supercritical Hopf bifurcation. Here, a focus loses stability and a stable T_0-periodic orbit is born. Fig. 4.2 sketches the phasespace before (a) and after (b) the bifurcation. Before the bifurcation, a stable focus exists (black dot in Fig. 4.2 (a)). A system trajectory (arrowed line) moves towards the steady state. After the bifurcation, a stable limit cycle (circle) surrounds the now unstable focus. Trajectories approach the limit cycle, and depart from the steady state. In such a point of operation, the unstable focus shall be stabilized by noninvasive control, and the stable oscillations along the limit cycle shall be suppressed.

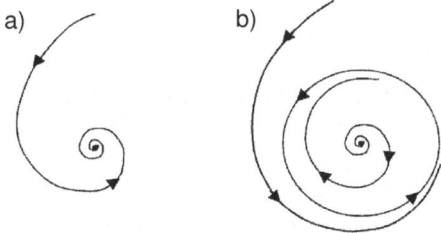

Figure 4.2: *Supercritical Hopf bifurcation in 2-dim phase space. a) Stable focus before (black dot) and (b) stable periodic orbit (circle) after the bifurcation. System trajectories (arrowed lines) approach the stable focus in (a) and the stable limit cycle in (b), respectively.*

Experimentally, a Hopf bifurcation in the ITL is recognized by a characteristic behavior of the noise-induced precursor previous to the bifurcation, and by the occurrence of undamped relaxation oscillations (RO) after the bifurcation. Bifurcation parameter is the current I_P in the passive laser section, which mainly changes the internal coupling phase between the two lasers in the ITL (Sec. 2.1).

When approaching the Hopf bifurcation, the damping of the damped RO decreases linearly until it crosses zero at the Hopf point. Before the crossing, the focus is still stable. However, the only slightly damped RO are pushed by noise, and a weak precursor appears in the power spectrum at $1/T_0 \approx 12$ GHz (gray line in Fig. 4.4 (a)). The width Δv of this peak is proportional to the RO damping. Prior to the bifurcation, Δv decreases linearly (open circles in Fig. 4.3). The RO amplitude is still low here (black squares). Beyond the bifurcation, the RO amplitude starts to grow rapidly, while the width stays on a minimum value. The laser has

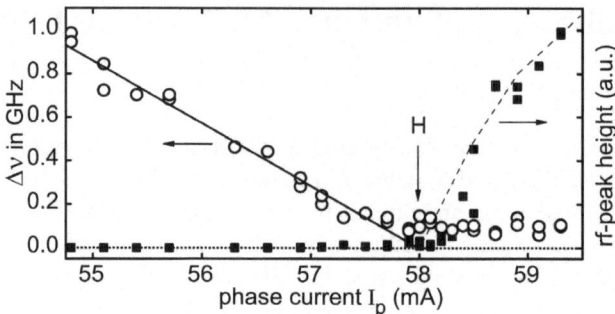

Figure 4.3: *Supercritical Hopf bifurcation in the freerunning ITL. Width (circles) and height (squares) of the dominant peak in the power spectrum of the solitary ITL versus phase current. Dashed: guide for the eye. Solid: linear fit for determining the Hopf bifurcation (H). DFB currents: $I_1 = 30$ mA, $I_2 = 45$ mA.*

changed to a stable selfpulsation, and the focus is now unstable. A linear fit to the data of the peak width (full line) yields the bifurcation point H at $I_P = 58.1$ mA.

Coupling of the Fabry-Perot Etalon

The unstable focus behind the Hopf bifurcation shall be stabilized by all-optical DFC. As discussed above, the delay τ should optimally be half the intrinsic focus period [HS05]. Half-period delay has also proven useful when increasing the stability of a stable focus in Sec. 3.1. The RO frequency at the Hopf bifurcation is about $1/T_0 \approx 12$ GHz. Thus, a FP etalon with $1/\tau \approx 26$ GHz roundtrip frequency is selected (Fig. 3.2).

In this experiment, an unstable stationary state must be adjusted resonantly to the FP. This is simplified by tracing the resonantly adjusted stable focus through the bifurcation. The ITL is operated in a point closely before the Hopf bifurcation, and the cw emission is adjusted resonantly to the FP. Then, I_P is increased stepwise and the laser is driven through the Hopf bifurcation. Increasing I_P also affects the emission wavelength, and this way deteriorates the resonant adjustment. Thus, each current shift affords a slight accommodation of the temperature in order to maintain resonance. The necessary overall temperature variation is far below 1 K. The essential characteristics of the Hopf bifurcation are not affected by the applied temperature corrections. The Hopf bifurcation is crossed along a line of constant wavelength in the current-temperature plane.

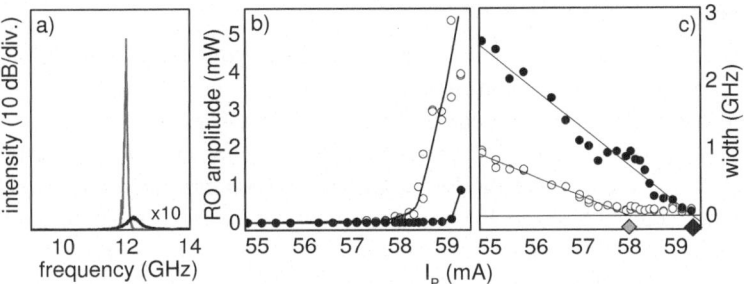

Figure 4.4: *Control of unstable focus. (a): Powerspectra of freerunning (gray) and controlled laser (black). The spectrum under control is multiplied by a factor of 10 for better visibility. (b), (c): Tracing of the stabilized focus versus bifurcation parameter I_P. RO amplitude (b) and peak width (c) versus I_P. Open symbols: freerunning laser, black dots: laser with FP. Black lines: (b): guide for the eye, (c): linear fit. FP parameters: $1/\tau \approx 26$ GHz, $R = 0.76$.*

Tracing the Unstable Focus

In this way, the focus could indeed be followed through the bifurcation. With resonant FP feedback, the Hopf bifurcation is 'shifted' by $\Delta I_P = 1.4$ mA from $I_P = 58.1$ mA to $I_P = 59.5$ mA[1]. Within this range, the unstable focus is controlled against the freerunning stable RO. Fig. 4.4 (a) compares measured powerspectra of the freerunning and the controlled laser. The RO amplitude is decreased by two orders of magnitude in presence of the FP. The frequency deviation between stabilized focus and freerunning RO is indicative of the fact, that the stabilized state is different from the freerunning state before the bifurcation. The RO peak width with FP, and thus the RO damping, is clearly above the corresponding freerunning values (Fig. 4.4 c)). Prior to the bifurcation, the RO damping of the stable focus is enlarged, in accordance with the experiments in Sec. 3.1. In all measurements, the latency phase φ was adjusted to an optimal value. The latency time τ_l and the feedback strength K are set to $\tau_l \approx 3\tau$ and $K \approx 0.05$, respectively. Both values have been proven useful in preceding simulations.

Impact of the Optical Phase

The effect of φ is studied in a fixed point of operation within the control range. φ is tuned over several periods by shifting the FP etalon towards the laser in small

---

[1]This is not a shift in the sense of a shift of the point of operation, see also the next section.

steps. This yields cyclic variations of both the RO peak height and the relative FP transmission (Fig. 4.5 (a), (b)). Control is attained within recurrent regions characterized by low RO amplitude and maximal relative transmission. The remaining reflected control signal is estimated to be less than 1 per mille of the laser intensity. The control therefore clearly has noninvasive character. Regions of control change with regions, where the RO amplitude is even enhanced compared to the freerunning state, while the FP transmission is low. Here, nonresonant Fourier components reduce the transmission signal. The nonvanishing FP feedback leads to more coherent selfpulsations. The observed behavoir is in excellent accordance with results obtained in accompanying numerical simulations (Fig. 4.5 (c), (d)).

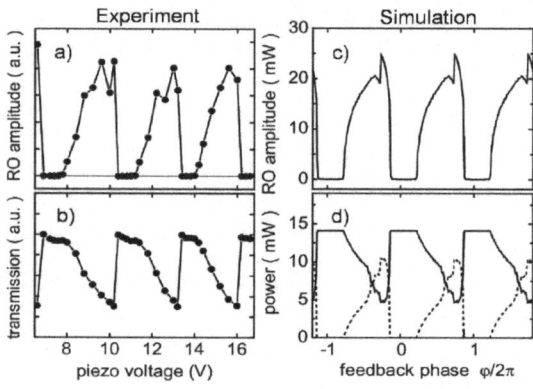

Figure 4.5: *Optical DFC of an unstable focus versus the latency phase φ. (a), (b) Experiment. (a) RO amplitude versus the piezo voltage (corresponding to φ). (b) relative transmission of the FP. (c), (d) Simulations. (c): RO amplitude and (d): FP transmission (solid) and reflected signal (dotted) vs φ.*

4.2 Investigation of Control Parameter Space

The control measurements are preceded by numerical simulations to identify suitable values for the control parameters. The calculations reveal a deeper understanding of the conditions for optical DFC, and confirm the experimental results.

Noninvasive Control beyond a Hopf Bifurcation

First, a supercritical Hopf bifurcation is prepared in the simulated ITL. In the simulations, the bifurcation parameter is the internal phase shift $\varphi_p = 2L_p \delta_p$ in

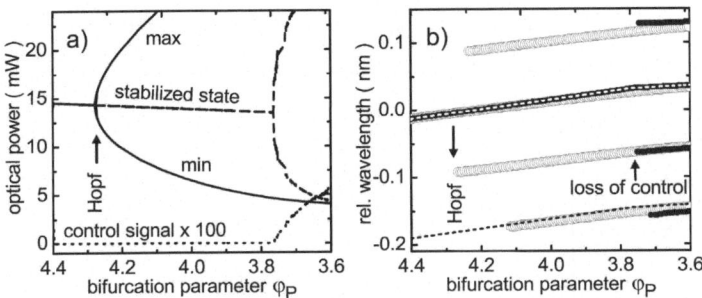

Figure 4.6: *Optical DFC in numerical simulations. (a): Maximum and minimum of the laser emission vs. internal phase shift φ_p. Solid: Solitary ITL, Dashed: ITL with lossless FP ($\varphi = 0, K = 0.1, e^{i\phi} = 1, \tau = 45$ ps $\approx T_0/2, R = 0.7, \tau_l = \tau$). Dotted: Time averaged control signal E_b. (b): main peaks in the optical spectrum (within 20 dB below maximum). Gray circles: Solitary ITL, thick black: ITL with lossless FP, dashed: resonances of the FP. ITL currents: $I_1 = 8$ mA, $I_2 = 80$ mA, $I_P = 0$ mA.*

the passive section. Note that increasing I_p corresponds to decreasing δ_p in experiment. Fig. 4.6 (a) shows the minimum and maximum value of the emitted power taken from transients calculated over a time interval of 8 ns. Equal values mean cw output. The solitary ITL undergoes the Hopf bifurcation under study at $\varphi_p = 4.277$. In full analogy to the experiment, relaxation oscillations with a period of $T_0 \approx 87$ ps become undamped here. The bifurcation is also evidenced by the optical spectrum, where extra peaks equidistantly separated by the frequency of the selfpulsations emerge (Fig. 4.6 (b)).

With latency and resonant FP section added, cw operation is maintained beyond the solitary bifurcation point. The feedback signal practically disappears in the whole stabilization range, confirming the noninvasive character of the control. Only a single peak is present in the optical spectrum. Like in the measurements, the peak shifts under variation of the bifurcation parameter. Thus, the resonance frequencies of the FP cavity have to be readjusted. An iteration procedure is applied that stops when at least one spectral line coincides with a FP resonance. Control is lost at $\varphi_p = 3.765$. Similar to the measurements, the bifurcation point is 'shifted' with FP feedback. However, the disappearance of the control signal (dotted line in Fig. 4.6 (a)) confirms that this shift is not simply a shift of the point of operation which may be caused by a finite feedback from the FP. The loss of control at $\varphi_P = 3.765$ is due to a Hopf bifurcation. This represents a common way to lose control in DFC when crossing a border of the control domain[2].

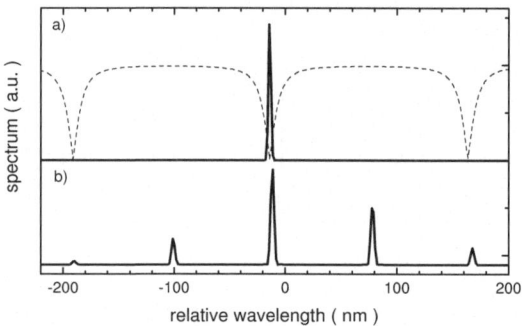

Figure 4.7: *Control of an unstable focus in the optical domain. (a) Optical spectrum of the stabilized cw state (thick solid line) plotted with the reflectivity spectrum of the FP (dotted line). (b) Optical spectrum of the freerunning selfpulsation in comparison.*

In what follows, a fixed point of operation ($\varphi_p = 4.15$) beyond the freerunning bifurcation point is considered. Here, the freerunning ITL is selfpulsating (Fig. 4.7 (b)). When adding the FP (dotted line in Fig. 4.7 (a)) with a feedback strength of $K = 0.05$, the laser changes to cw emission (thick line in Fig. 4.7 (a)). The controlled cw emission sits exactly in a FP reflectivity minimum, and is slightly shifted compared to the central optical line of the freerunning SP (Fig. 4.7 (b)).

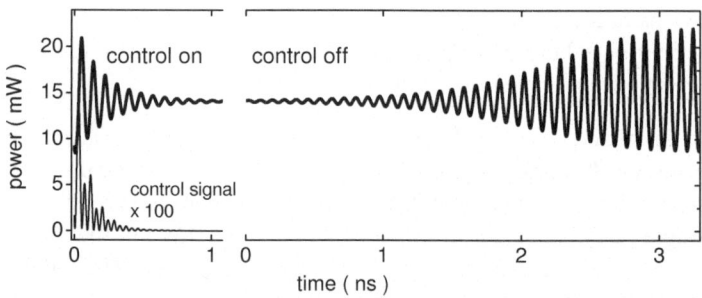

Figure 4.8: *Transient of the emitted output when switching control on (a) and off (b). Thick line: emitted optical power, thin line: control signal, multiplied by 100 for better visibility. Parameters of Fig. 4.6 and $\varphi_P = 4.15$.*

---

[2]The bifurcations occurring at the control domain borders in optical DFC are studied also in Chapter 6.

Control Dynamics

The simulations provide insight into the dynamics of the control force that is hardly obtainable experimentally. When 'switching control on' by setting K from zero to $K = 0.05$, the ITL operation turns from a freerunning stable SP into the controlled stationary regime. The timetrace of the emitted power in the left part of Fig. 4.8 shows this transition (thick solid line). The control signal returned from the FP section is initially about 1% of the 15 mW device output, but it drops dramatically down. The steady state is approached with an exponential time constant of $\tau_{con} \approx$ 260 ps. This time is a quantitative measure of the control efficiency. Under the reverse switch of K, the selfpulsation recovers (right panel of Fig. 4.8). The rise time of 0.59 ns and the frequency of 11.5 GHz of the small-amplitude oscillations arising at the initial stage represent the complex eigenvalue $\lambda - i\omega$ of the unstable focus. The frequency 11.2 GHz of the fully developed pulsation is only slightly slower, because the point of operation is still close to the Hopf bifurcation.

Domains of Control

In the following, the domain of control is investigated for the control parameters K, φ, and τ_l. One-parameter variation of only φ, with K kept fixed, yields a behavior similar to the experiment. Fig. 4.5 shows the effect of φ on the RO amplitude (c), and on the power transmitted and reflected by the FP (d). A periodic exchange between control and destabilization of the focus, analogous to the experimental data, is observed. Two-parameter control domains are calculated as follows. The initial state of the device is always set to a selfpulsation of the freerunning ITL, and then control is switched on. The criterion of noninvasive control is that the emitted power and wavelength asymptotically approach the values of the unstable focus for $K = 0$.

Fig. 4.9 depicts the control islands in the (φ, K) plane. Within the black regions, noninvasive control of the unstable focus is achieved. Control requires a minimal feedback strength $K = 0.02$, and is lost again for too high K, when nonresonant Fourier components are induced by the feedback. As already noted, control is only possible in a finite range of φ. The maximal range in φ is found for moderate feedback strengths around $K = 0.05$. Within the gray regions in Fig. 4.9, the laser is also operating in a cw state, but nonresonant to the FP. Here, a finite feedback signal arises from the FP.

Successful control further depends on the latency time τ_l. From generic models [HS03] it is known that DFC only works in a finite range of τ_l, and fails for too high latency times. However, the numerical simulations refined the restriction for

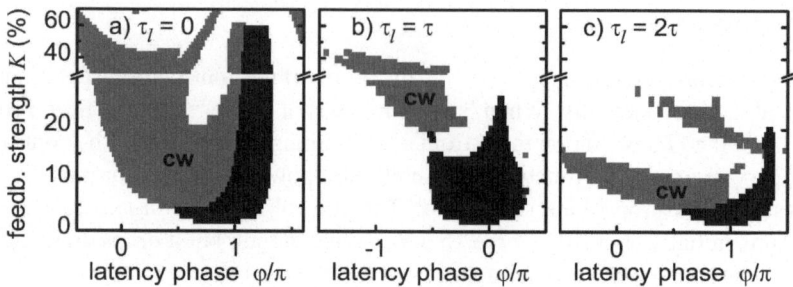

Figure 4.9: *Control domains in* (φ, K) *plane for different latency times* τ_l*. Black: noninvasive control. Gray: invasive, cw emission. White: no cw emission achieved.*

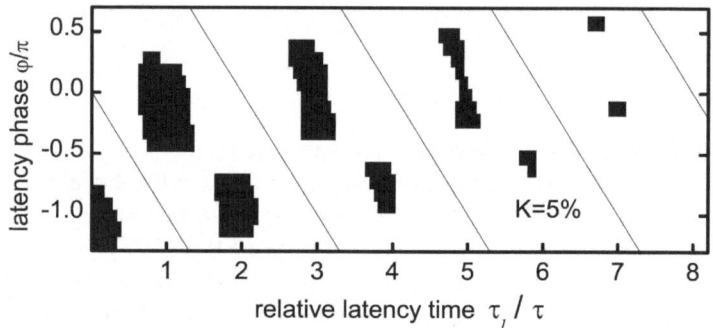

Figure 4.10: *Control domains in* (φ, τ_l) *plane for* $K = 0.05$*. Black: noninvasive control.* $\tau = T/2$.

τ_l for the all-optical setup. Evaluation of the domain of control in the (τ_l, φ) plane reveals a cyclic behavior (Fig. 4.10). Several control domains are located around integer values of τ_l/τ. The size of the control domains decreases with increasing τ_l, control is lost beyond $\tau_l/\tau \approx 7$. The control measurements in Sec. 4.1 are performed in the control island at $\tau_l/\tau = 3$. All other control measurements in this thesis operate in the $\tau_l/\tau = 1$ island. A direct experimental verification of the $\tau_l - \varphi$ relation was not possible, because the absolute value of φ is unknown after having readjusted τ_l. However, the measurements confirm optimum control near to integer τ_l/τ and loss of control for large τ_l.

4.3 Summary

The all-optical control setup is successfully used to control an unstable stationary state in the ITL. Resonant FP feedback is applied to a regime of stable selfpulsations shortly beyond a Hopf bifurcation. Under control, the laser shows stable cw emission, accompanied by minimal feedback from the FP. Thus, the unstable stationary state is recovered noninvasively. The RO could be suppressed by two orders of magnitude. Control of the stationary state is achieved within a range of $\Delta I_P = 1.4$ mA beyond the bifurcation. As expected, successful control is sensitively dependent on the latency phase φ.

Accompanying numerical simulations confirm the experimental results and additionally allow for a study of the transient dynamics and of the domains of control. Particularly, the influence of the latency time τ_l is evaluated. The general analytic result, that τ_l must be kept as small as possible is refined here.

In this configuration, the fraction τ_l/τ has to be approximately integer in order to achieve control. Taking this into account, τ_l is set equal to τ in all following experiments.

The results of this first all-optical control experiment are published in [SHW$^+$06] and [WSH08].

5 Control of Unstable Selfpulsations

After extensive preparation and a first control experiment with unstable stationary states, the all-optical setup is used to stabilize an unstable periodic orbit in the MSL. DFC of unstable periodic orbits (UPOs) has been experimentally realized in various systems in physics, chemistry, biology, and medicine [Sch99], as well as in economics (see, e.g. [HU00]). The first control experiments were performed with electronic circuits [PT93, GSCS94]. Up to now, DFC has been experimentally proved with chaotic lasers [BDG94, DTH98], a magneto-elastic beam [WYP95], gas discharges [PBA96], chemical reactions [PMR$^+$99b, BBM$^+$03], airfoils [RN01], or a Taylor-Coutte flow [LWP01]. The fastest controlled pulsations reported so far operate on timescales of 100 MHz frequency, and are controlled by means of optoelectronic feedback [BIG04]. In what follows, the control of an UPO with 26 GHz frequency is demonstrated.

In the present context, an UPO corresponds to an unstable selfpulsation (SP) of the emitted amplitude.

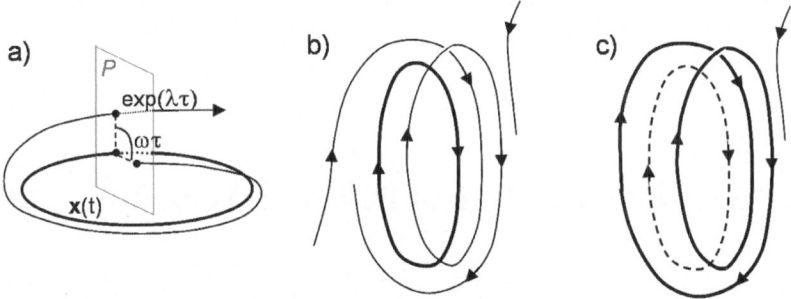

Figure 5.1: *(a) Trajectory in vicinity of UPO x(t) with period T. P: Poincare plane. (b), (c) Period doubling bifurcation in 2-dim projection of the phase space. (b) Stable period-1 orbit (thick solid) before and (c) stable period-2 orbit (thick solid) and unstable period-1 orbit (dashed) after the bifurcation. Thin lines: neighboring system trajectories.*

The local stability of a periodic orbit can be analyzed by a linear stability analysis, analogous to the procedure in Chapter 4 for a fixed point. The local stability of a periodic orbit in an n dimensional system is governed by n characteristic multipliers with complex Floquet exponents $\Lambda_{0,k} = \lambda_k \pm i\omega_k$. One of them is always zero, and refers to deviations along the orbit trajectory itself. The remaining $n - 1$ exponents can be analyzed in a projection of the full n-dimensional problem to a surface of section, which is normal to the orbit trajectory, the Poincare plane (see Fig. 5.1 (a)). The results derived for stationary states can be transfered to this $n - 1$-dimensional map [Ott].

The orbit is unstable if at least one of the remaining $n - 1$ exponents has a real part $\lambda_k > 0$. Then, small deviations are amplified and system trajectories eventually depart from the orbit. The corresponding value of ω_k describes the torsion of a departing trajectory. For finite $\omega_k > 0$, a trajectory leaving the UPO rotates by an angle $\omega_k T$ per period T about the orbit (Fig. 5.1 (a)). The case of a real positive $\Lambda_{0,k} = \lambda_k > 0$ with $\omega_k = 0$ is called a torsionfree unstable orbit. Stabilization of a torsionfree UPO by DFC is impeded by the odd-number limitation (Chapter 7). In contrast, maximal torsion $\omega_k = \pi/T$ presents optimal conditions for DFC. Simplified, in this case the control force is maximal thanks to the contribution of a finite ω_k [JBO$^+$97b]. Unstable orbits generated in period doubling bifurcations show maximal torsion, they 'flip their environment' in one period.

Thus, in the following experiments an UPO shortly beyond a period doubling bifurcation is adressed. In a period doubling bifurcation, a periodic orbit loses stability in favor of an orbit with double period (see Fig. 5.1 (b) and (c)). This type of bifurcation is prevalent in dynamical systems, and also appears in the studied MSL devices. A cascade of subsequent period doubling bifurcations is one of the basic routes to chaos observed in dynamical systems.

5.1 Experimental Optical DFC of a Periodic State

In the specific ITL device under study (Sec. 2.1), the chaotic regime is mostly approached via period doubling routes to chaos. This scenario is stable in a large range of the laser parameters. Thus, an independent tuning of both the pulsation frequency $1/T$ and of the central emission frequency ω_0 is possible without leaving the dynamic regime. The resonant coupling of a SP to the FP etalon requires the simultaneous adjustment of four parameters: ω_0 is tuned via the device temperature in order to fulfil the resonance condition (2.3), one of the laser currents defines the pulsation period T to meet the delay condition (2.2), the latency phase φ must be adjusted to a suitable range, and the current I_P acts as bifurcation parameter.

The Freerunning Period Doubling Bifurcation

The ITL is operated along a period doubling route to chaos as shown in Fig. 5.2. Initially, the laser shows stable SP, recognized by a sharp line in powerspectrum

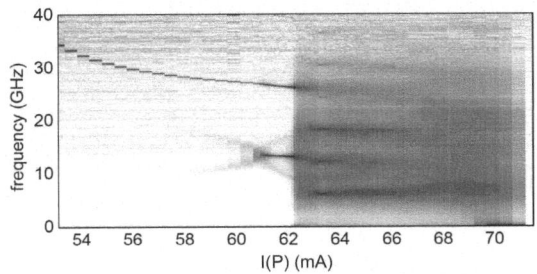

Figure 5.2: *Period doubling route to chaos in the freerunning ITL. Powerspectra of the emitted intensity are plotted in logarithmic grayscale versus the vertical axis for different values of the current I_P on the horizontal axis. $I_1 = 47.8$ mA, $I_2 = 74$ mA, $T = 20°C$.*

at $1/T = 34$ GHz. The pulsation frequency decreases with increasing bifurcation parameter I_P. At $I_P = 61.3$ mA, an additional line at half the frequency appears in powerspectrum: a period doubling bifurcation occurs. Here, the stable SP (in following called 'period-1 SP') loses stability in favor of a SP with double period. On further increasing I_P, the laser quickly enters chaos. A true 'cascade' of period doubling bifurcations is not observed here. In real systems, noise mostly prevents the observance of a full period doubling cascade to chaos. However, three facts indicate that this transition is indeed a period doubling bifurcation: the observance of a virtual Hopf precursor [Wie85] prior to the bifurcation, a period-4 window in chaos around $I_P = 63$ mA, and the occurrence of a second period doubling in neighboring points of operation.

Control of the Unstable Period-1 Selfpulsation

The aim is the stabilization of the period-1 SP behind the period doubling bifurcation. The frequency of the period-1 SP closely before the bifurcation point is $1/T \approx 26$ GHz. A FP etalon with $1/\tau = 26.1$ GHz roundtrip frequency is added. First, the laser is operated shortly before the bifurcation, and the still stable period-1 SP is carefully adjusted to resonance with the FP. It is searched for a control parameter combination (K, φ) where the coherence of the stable period-1 SP is improved, and the FP transmission is maximal. This stage of 'coherence control' (Sec. 3.2) presents a preliminar stage to the control of an unstable orbit. Then,

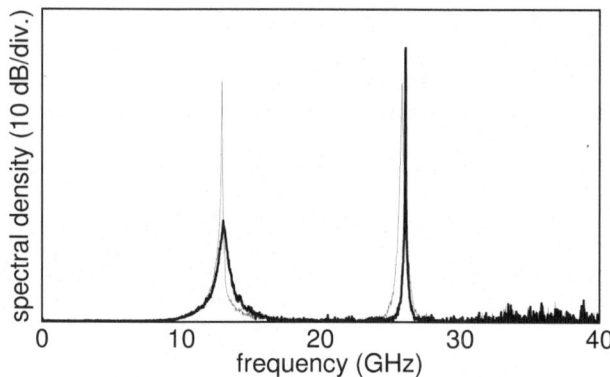

Figure 5.3: *Control of the period-1 pulsation. Powerspectra of the freerunning laser (thin gray line) and of the controlled state (thick black line).* $(I_1, I_P, I_2) = (47.8, 67.8, 72)$ *mA, T* $= 20.3° C$. *FP parameters:* $R = 0.76, 1/\tau = 26$ *GHz, K=0.04,* $\tau_l \approx \tau$. *Please note the logarithmic scale.*

the obtained values for K and φ are held fixed, and the bifurcation is passed by increasing I_P stepwise. For each measuring point, resonance of the emission with the FP has to be restored via small adjustments of temperature and laser current.

This way, the period-1 SP can indeed be followed successfully through the bifurcation. Behind the bifurcation, the period-1 SP remains stable under control, and a strong suppression of the period-2 SP around 13 GHz by more than two orders of magnitude is achieved (please note the log scale in Fig. 5.3). The stabilized period-1 SP could be followed by $\Delta I_P = 0.7$ mA behind the bifurcation.

Dependence of Control on Latency Phase

Variation of φ yields the expected periodic exchange between regions of stabilization and regions of destabilization (Fig. 5.4 and Fig. 5.5). Three regions of control of the period-1 SP are crossed. Off the control regions, the period-2 SP are enhanced by a nonvanishing feedback from the FP. Fig. 5.4 shows cutouts of measured powerspectra around the period-1 peak (middle) and around the period-2 peak (left) for different values of the piezo voltage. The right panel plots the corresponding relative transmission of the FP. Under control, the relative transmission of the FP is maximal, indicating minimal control power, and the period-1 frequency is constant. Additionally, the line in powerspectrum at the period-2 frequency is suppressed. Thus, in the control regions an otherwise unstable SP in

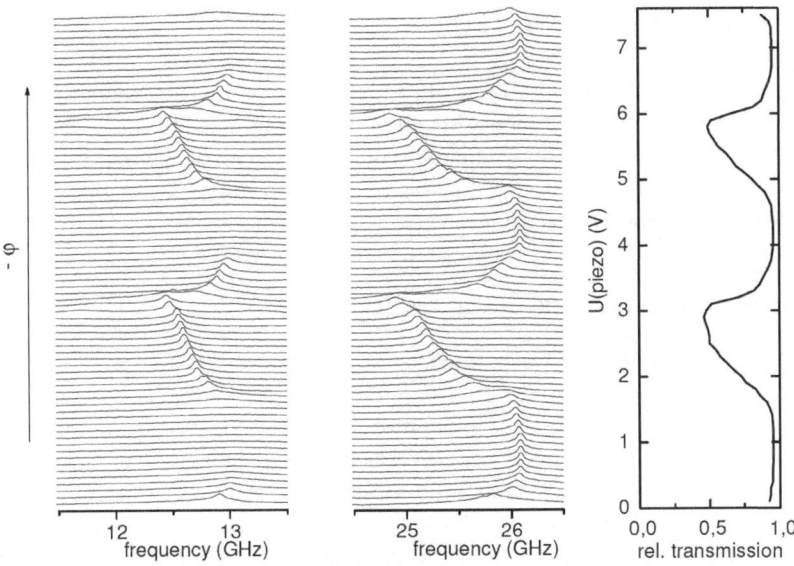

Figure 5.4: *Phase-dependent control of the period-1 pulsation. The latency phase φ decreases in vertical direction. Left and middle panel: Cutouts of the powerspectra at the period-2 and at the period-1 line, respectively. Powerspectra are plotted in logarithmic scale. Right panel: Relative transmission. Same point of operation as in Fig. 5.3.*

the ITL can be observed directly in the measured powerspectra. The period-1 pulsation frequency is constant under control at a value of 26.1 GHz (Fig. 5.5 (c)), slightly above the freerunning frequency component at 25.8 GHz (dotted line). Latter value corresponds to the second harmonic of the freerunning period-2 SP. In the regions of control, the amplitude of the period-2 SP is suppressed strongly below the freerunning value (Fig. 5.5 (a)), while the damping is strongly increased (Fig. 5.5 (b)). Additional evaluation of measured optical spectra reveals that also the optical frequency is constant under control (Fig. 5.6 top). The control regions between $U_{piezo} = 0...1$ V and $U_{piezo} = 3.6...4.6$ V are identified by a maximal relative transmission (bottom).

The constancy of both optical and pulsation frequency within the regions of control, accompanied by maximal relative FP transmission, indicates that the stabilized period-1 SP is a genuine state of the laser dynamics. The unstable SP is controlled noninvasively.

5.2 Domains of Control

The control of the period-1 SP by all-optical DFC is repeated in the simulated ITL. The calculations yield deeper insight into the conditions for the control of the period-1 UPO within chaos. In the experiment, a tracing of the period-1 SP into chaos is not achieved. However, the ideal conditions in the calculations enable a tracing of the period-1 SP deep into the chaotic region.

Bifurcation Sequence without and with Control

In the simulated ITL, PD routes to chaos are also prevalent. Fig. 5.7 (a) shows a typical bifurcation sequence of the freerunning device. Here, two parameters are increased simultaneously, the internal phase shift φ_P in the passive section and the laser current I_1. This procedure yields a constant frequency of the period-1 SP along the bifurcation sequence, similar to the experiment. The freerunning laser shows a cascade of 5 successive period doubling bifurcations, followed by the onset of chaos (Fig. 5.7 (a)). The first period doubling occurs at $\varphi_P = -0.068$. A period-1 SP with a frequency of $1/T = 28$ GHz loses stability, and a period-2 SP with a frequency of $1/T = 14$ GHz takes over. Closely behind this bifurcation, a

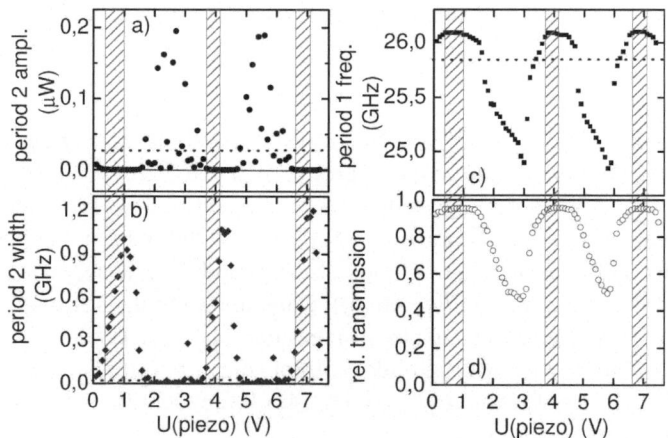

Figure 5.5: *Control of the period-1 UPO versus φ. (a),(b) Amplitude and width of the period-2 peak, respectively, (c) frequency of the period-1 peak and (d) relative transmission versus the piezo voltage. Dotted lines in (a)-(c): freerunning values. Shaded areas: Control of the period-1 pulsation. Same point of operation as in Fig. 5.3.*

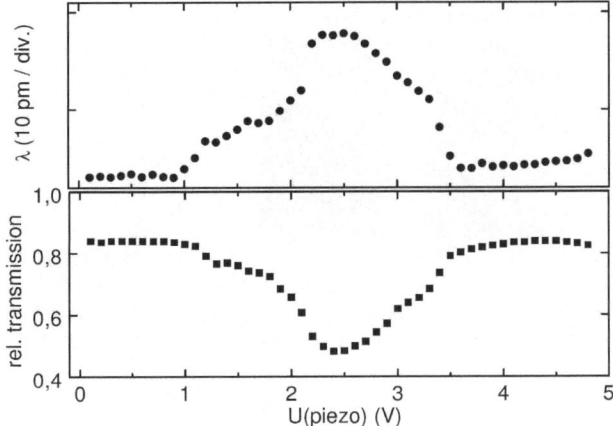

Figure 5.6: *Phase-dependence of the emission wavelength under control. Top: Wavelength* λ *of the main emitted optical mode and bottom: relative transmission versus piezo voltage. Same point of operation as in Fig. 5.3.*

FP section with τ equal to T is added to the laser. The control parameters K and φ are varied until stabilization of the period-1 SP is achieved. Then, K and φ are fixed, and the bifurcation sequence is followed by increasing φ_P and I_1. Under the ideal noisefree conditions in the simulations, the stabilized period-1 SP is traced deep into chaos (Fig. 5.7 (b)).

Domains of Control

A comparison of domains of control at the bifurcation ($\varphi_P = -0.068$) and in chaos ($\varphi_P = 0.0$) explains why the stabilization of the period-1 SP within chaos could not be achieved experimentally. The control domains in (φ, K) plane (Fig. 5.7 (c) and (d)) are calculated as follows. For each parameter pair (K, φ) the laser is first set to a freerunning chaotic state. Then, control is 'switched on'. Successful stabilization is identified by the occurence of a pulsation with period τ, accompanied by the disappearance of the control signal after a short transient. Due to the finite integration time of the computations the control signal does not reach zero, but a finite control power remains. Black regions denote a final control signal below 10^{-5} of the level of the laser emission. As in the experiment, control is limited to a certain range of the latency phase φ. For all suitable φ, there is a lower and an upper border for the feedback strength K. At the bifurcation (Fig. 5.7 (c)), the

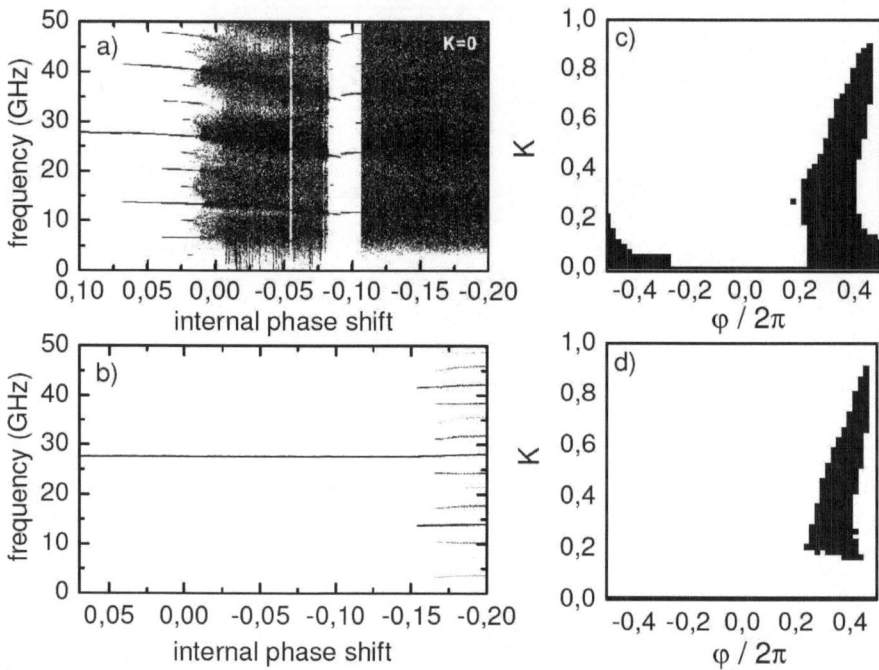

Figure 5.7: *Control of the period-1 SP in simulations. (a) Bifurcation sequence in the freerunning laser when tuning φ_P and I_1 simultaneously. (b) The corresponding sequence with coupled FP. τ matches the period of the freerunning period-1 pulsation at $\varphi_P = -0.068$. (c), (d) Domains of control in (φ, K) plane shortly after the bifurcation at $\varphi_P = -0.068$ and in chaos ($\varphi_P = 0.0$), respectively.*

lower border is at $K_{min} = 0.01$, corresponding to the grid size of K in the calculations. In chaos (Fig. 5.7 (c)), the lower border rises to about $K_{min} \approx 0.2$. K_{min} rises even more when going deeper into the chaotic region. In the experiment presented in the previous section, a much lower feedback strength $K = 0.04$ is used. A feedback strength of the order of $K_{min} \approx 0.2$ would probably lead to feedback-induced destabilization of the laser, and is thus not convenient.

In above simulations, parameters similar to the values in table 2.3 are used, except for $\tau_l = 127$ fs, $\tau = 35973$ fs, $I_1 = 100.4$ mA, $\delta_1 = 19341$ m^{-1}, $L_P = 500$ μm, $\gamma = 25$ cm^{-1}, $I_2 = 80$ mA, and $\delta_2 = 16321$ m^{-1}.

5.3 Summary

The all-optical DFC setup is applied successfully to control an unstable selfpulsation in the ITL. An unstable period-1 SP, which has been destabilized in a period doubling bifurcation, could be recovered noninvasively. Under control, optical frequency and pulsation frequency are constant within a finite range versus the latency phase φ, accompanied by a maximum of the power transmitted by the FP. This strongly indicates the noninvasive nature of the achieved control. The RO amplitude of the period-2 SP, which gets undamped in the freerunning bifurcation, could be suppressed by more than two orders of magnitude.

Numerical simulations give a deeper insight into the control parameter range. Under ideal noisefree conditions, a stabilization of the unstable period-1 SP shortly after the bifurcation is possible within a huge domain of control in (φ, K) plane. The unstable period-1 SP could be stabilized also within the chaotic region. But here, feedback strengths above $K = 0.2$ are required. This value is not convenient for the experiments, where feedback-induced destabilization of the laser must be avoided. Thus, in the following measurements it is refrained from stabilizing the period-1 SP in chaos. Instead, chaos control is tested for the unstable period-2 SP in chaos. It is expected, that in this case control will be possible for lower K.

6 Controlling Chaos

In this thesis, all-optical control experiments are realized in dynamical regimes of increasing complexity, following a 'virtual' route to chaos like the one in Fig. 2.6. Finally, in this Chapter chaos is reached.

The concept of chaos control is especially interesting in conjunction with lasers. Chaos control in lasers means that an irregular emission is transformed to periodic intensity pulsations or stationary emission. Control of chaotic lasers has been demonstrated by different feedback schemes and non-feedback methods. Nonfeedback methods of chaos suppression involve the application of small driving forces, i.e. weak periodic perturbations to a control parameter [MGA94]. The period of the perturbation is conveniently chosen equal to a period of an existing UPO, but also higher harmonic [CMA95] and subharmonic [VFB95] perturbations are reported. Of course, this method acts invasive: the dynamics are slightly modified such that stable solutions appear. The main advantage of nonfeedback methods is that they are fast, involving no real-time monitoring and computing.

The two possible diametric effects of feedback - stabilization or destabilization - are also well-known for lasers [LK80]. Even small feedback to a laser often acts destabilizing, causing a regularly operating laser to emit irregular fluctuations of low frequency [RV77], or inducing coherence collapse [LVdB85] of the laser. On the other side, appropriate feedback can decrease the linewidth of the laser emission [POS+83], or render an irregular emitting laser stable. The feedback methods used to control laser chaos are the OGY method and its implementations 'occasional proportional feedback' [RMM+92] and 'minimal expected deviation' [RFBB93], electronic and optoelectronic DFC [BDG94, DTH98, BIG04], and the method of negative feedback of subharmonic components [MCA96], which has strong analogies to DFC.

All these control schemes involve electronic loops. Thus, they are restricted in speed. So far, maximal frequencies of 100 MHz could be successfully stabilized [BIG04]. In high-speed data communication, multisection semiconductor lasers operate in the 10-GHz range and beyond. These ultrafast timescales demand all-optical feedback. Chaos control of MSL is not reported so far, even though this method could be especially interesting in conjunction with fast communication schemes. This gap is closed with the proof-of-concept experiments presented in this Chapter.

I achieved the first control of chaos by purely optical means. The stabilized pulsation represents the fastest dynamics which could be controlled so far. A 13 GHz periodic selfpulsation in a multisection semiconductor laser is stabilized in an otherwise chaotic point of operation. The achieved control is both robust against small parameter perturbations and has been found to be easily reproducable. Numerical experiments investigate the role of the control parameters. The results are published in [SWH08].

6.1 Chaos Control in Experiment

The control experiment in chaos also aims at an UPO born in a period-doubling bifurcation. Because the tracing of a period-1 SP into chaos failed (previous Chapter), a period-2 SP is adressed in the following.

Freerunning Route to Chaos

The freerunning ITL is operated along the route to chaos shown in Fig. 6.1 (a). First, the selfpulsating laser undergoes a period doubling at $I_P = 61.6$ mA. Here, the fundamental SP with $f = 25$ GHz frequency loses stability, and a period-2 SP with $f/2$ appears. At $I_P = 65$ mA, the laser emission becomes chaotic, characterized by a broad powerspectrum. The two powerspectra in Fig. 6.1 (c) and (d) show the rising of the chaotic background. At the onset of chaos, the period-2 peak has a frequency of $f = 13$ GHz (Fig. 6.1 (c)). A broad chaotic background is already visible. On going further into chaos, the chaotic background grows, while the period-2 peak collapses to a weak feature around 13 GHz (Fig. 6.1 (d)). This small remaining feature superimposed to the broad spectrum indicates the embedding of the period-2 SP as an unstable element within the chaotic attractor.

Control of an Unstable Pulsation in Chaos

The period-2 SP shall be recovered within chaos by all-optical control. Initially, the laser is operated shortly before the onset of chaos, where the period-2 SP is still stable. A FP etalon with 13 GHz roundtrip frequency is added to the laser. The selfpulsating laser emission is adjusted to resonance with the FP via adaptation of I_1, I_P, and the temperature. Then, control parameter values K and φ are searched which yield an increase of the coherence of the period-2 SP, accompanied by maximal FP transmission. The such determined values for K and φ are fixed in the following, and the laser is driven into chaos by increasing I_P. Slight simultaneous accommodations of the laser current I_1 and of the temperature are necessary

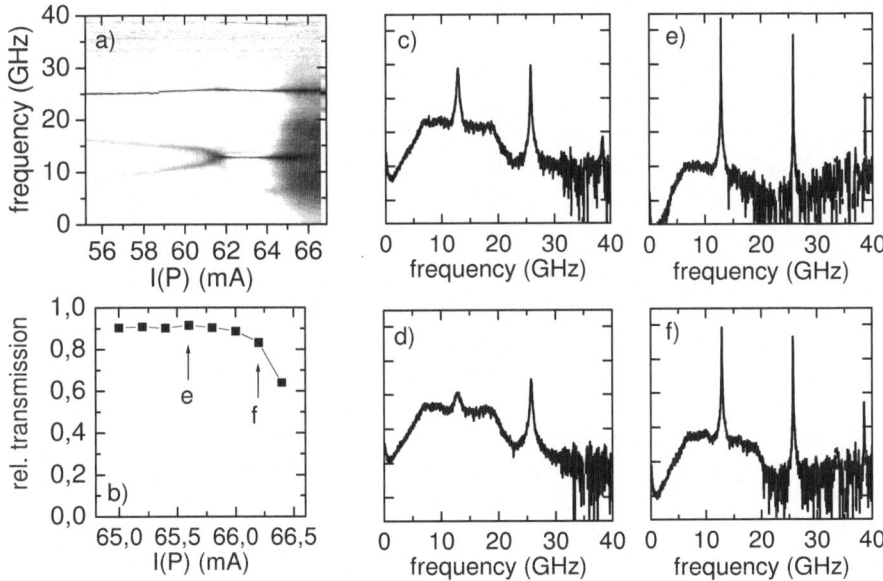

Figure 6.1: *All-optical control of chaos. Tracing of a period-2 orbit with $f = 13$ GHz into the chaotic region. (a) Route to chaos in the freerunning ITL. Powerspectra in grayscale (log scale) versus bifurcation parameter I_P. $I_1 = 57$ mA, $I_2 = 74.4$ mA. (b) Relative transmission of the resonantly coupled 13 GHz FP etalon. (c), (e) Powerspectra of the freerunning and the controlled laser, respectively, at $I_P = 65.6$ mA. (d), (f) Powerspectra without and with control, respectively, at $I_P = 66.2$ mA. Please note the log scale in (c)-(f). FP parameters: R=0.76, $1/\tau = 13$ GHz, K=0.03, $\tau_l = \tau$.*

to maintain resonance with the FP. This way, a stabilization of the period-2 SP within chaos is indeed achieved (Fig. 6.1 (e) and (f)). The amplitude of the controlled period-2 peak increases in height by two orders of magnitude, and the peak becomes much sharper. Thus, a nearly perfect periodic pulsation is recovered. Please note the logarithmic scale in Fig. 6.1 (c)-(f).

The period-2 SP is traced along a range of $\Delta I_P \approx 1.2$ mA into the chaotic region. The FP transmission is high under control (Fig. 6.1 (b)). At $I_P = 66.4$ mA, the control signal rises, and on further increase of I_P control is lost abruptly.

The achieved control has been found quite robust against unavoidable small parameter drifts over time. Additionally, control of the period-2 SP could be easily reproduced in other points of operation. With a pulsation timescale of about $T \approx 77$ ps this represents the fastest chaos ever controlled experimentally.

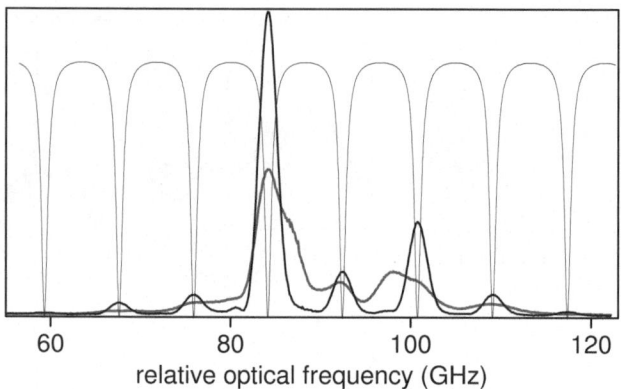

relative optical frequency (GHz)

Figure 6.2: *Control of chaos (optical spectrum). Measured optical spectra of the freerunning (gray) and of the controlled laser (thick black) plotted with the calculated reflectivity spectrum of the FP (thin black). Vertical axis arbitrarily scaled. Same point of operation as in Fig. 6.1 (f).*

Optical Spectrum

The transformation of the irregular emission to a regular pulsation can be nicely observed in the optical spectrum (Fig. 6.2). The spectrum of the freerunning chaotic laser shows rather broad structures with a pronounced carrier frequency (gray). Under control, the optical spectrum turns into a comb of equidistant lines (thick black). The peak width equals the resolution of the optical spectrometer. A calculated FP reflectivity spectrum is sketched by the thin black line in Fig. 6.2. The control leads to a reorganization of the optical Fourier components into the reflectivity minima of the FP interferometer.

Chaos Control versus φ

Above measurements are performed with optimized latency phase φ. Tuning of φ in a fixed point of operation leads to a periodic exchange of regions of control and regions where the diffusion in chaos is enlarged (Fig. 6.3 (a)). Latter regime is indicated by broader structures in powerspectrum, and accompanied by low FP transmission. Here, the laser receives strong feedback from the FP, which increases the irregularity of the chaotic state. The powerspectra are even less structured than in the uncontrolled device ((Fig. 6.3 (d)). Such behaviour might be interesting for applications demanding an extremely flat response. In contrast, the regions of

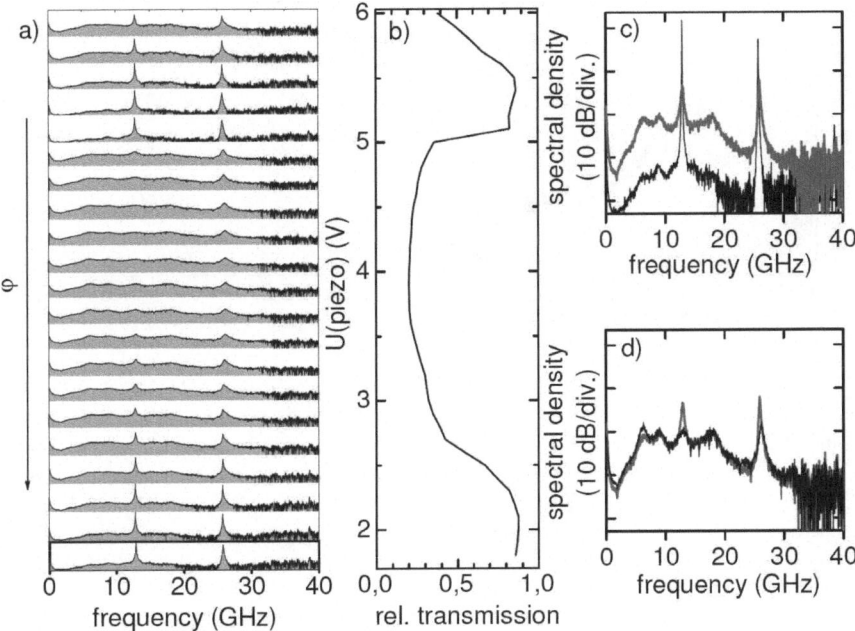

Figure 6.3: *Effect of the latency phase φ on optical chaos control. (a): Powerspectra obtained when decreasing φ in lateral direction, corresponding to selected piezovoltages in panel (b). (b): Relative power transmitted through the FP versus piezovoltage. (c), (d): Powerspectra with (black) and without FP (gray) at $U_{piezo} = 1.9$ V and 4 V, respectively.*

control are characterized by sharp peaks in powerspectrum, and by a high relative FP transmission (Fig. 6.3 (c)). Variation of the phase φ allows for a switching between these two regimes. This way, regular dynamics as well as even more irregular emission can be easily adjusted.

Noninvasivity of the Control

An important question is whether the achieved control is indeed noninvasive (see also Appendix C reviewing this aspect in literature). In a practical setup, the reflectivity at the FP minima remains finite, and noise is inevitably present. Thus, under control there is still a residual feedback signal entering the laser. An upper border for the remaining feedback is estimated from the level of the signal transmitted by the FP under control, and from the actual coupling strength of the FP. The residual

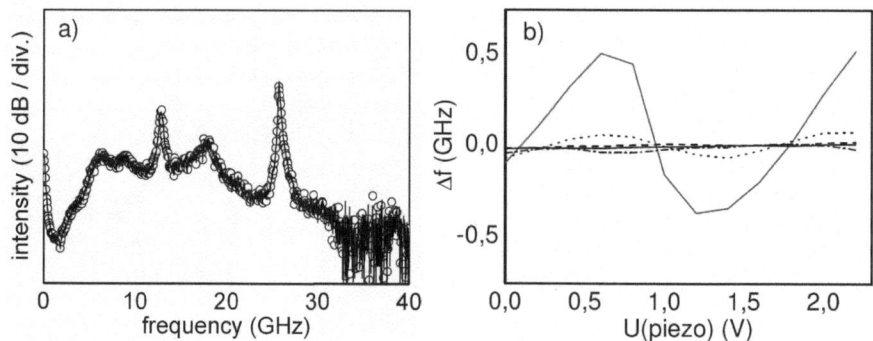

Figure 6.4: *Check of noninvasivity. a) Chaotic laser subject to mirror feedback. No change of operation mode is found when replacing the FP etalon by a simple mirror with the minimum FP reflectivity. Dots: powerspectrum of freerunning laser, solid line: powerspectrum with mirror feedback. b) Selfpulsating laser subject to mirror feedback. A stable 20 GHz selfpulsation receives feedback from a simple mirror. Deviation Δf of the pulsation frequency from the freerunning frequency versus the piezo voltage for different coupling strengths: $K_{mirror} = 0.016$ (solid gray), 0.012 (dot), 0.0024 (dash dot), 0.0005 (dash), 0.0001 (solid). Laser currents: $(I_1, I_P, I_2) = (37.7, 61, 25.5)$ mA.*

control signal is well below 10^{-3} of the laser emission. Though this value is rather small, the role of the remaining feedback is checked on in following experiment.

The FP is replaced by a simple mirror providing a feedback strength equal to the remaining control signal. In marked contrast to the FP setup, no change of the chaotic powerspectrum is observed (Fig. 6.4 (a)) for any value of the latency phase φ.

Further, the question is adressed how the remaining feedback affects the stabilized SP. The ITL is operated outside the chaotic regime where it shows stable SP, and feedback from a simple mirror is applied. The feedback strength is initially set to $K_{\mathrm{mirror}} \approx 0.016$. Tuning of φ in this regime yields an oscillation of the pulsation frequency around the freerunning value (gray line in Fig. 6.4 (b)). Then, the feedback strength is decreased stepwise to the level of the estimated maximal residual control signal. Thereby, the modulation depth of the frequency variation decreases to a level well below one per mille of the pulsation frequency. These findings provide conclusive evidence that the SP observed under control is not caused by a feedback-induced shift of the point of operation, but associated with an UPO existing in chaos, which is stabilized by noninvasive feedback from the FP.

6.2 Domains of Control

The control of an unstable period-1 SP in numerical simulations has been discussed already in Sec. 5.2. The period-1 pulsation could be traced along a complete period doubling cascade deep into the chaotic region (see again Fig. 5.7 (a), (b)). Domains of control at the period doubling bifurcation and within chaos (Fig. 5.7 (c), (d)) were derived. This section continues this investigation, providing a further analysis of the three-dimensional manifold of the control parameters K, φ and R within chaos. In the measurements discussed above, control of chaos is achieved for the unstable period-2 pulsation. In simulations, control of the fundamental period-1 pulsation, of the period-2 pulsation and of the period-4 pulsation is achieved within chaos. However, the following calculations focus on the conditions for control of a period-1 pulsation.

Route to Chaos in Simulations

The same point of operation as in Sec. 5.2 is considered. The ITL is driven into chaos along a period doubling route. Fig. 6.5 (a) shows the bifurcation sequence observed upon tuning of the internal phase shift φ_P. In all following calculations, the laser is operated at $\varphi_P = 0$, where the freerunning device is chaotic. The other parameters in these calculations are similar to the values used in Sec. 5.2.

Control Domain in (φ, K) Plane   In Sec. 5.2, a domain of control in (φ, K) plane is calculated at $\varphi_P = 0$ within chaos. Here, the domain is recalculated, this time also taking possible hysteretic behavior into account. The ITL is first set to the chaotic regime, and then the FP is added. Successful control is identified by the occurence of a τ-periodic SP, and by the disappearance of the feedback signal subsequent to a short switching transient. This way, the domain of control in Fig. 6.5 (b) is obtained. Here, the period-1 pulsation is locally stable. The chaotic initial state belongs to its basin of attraction only in the dark regions of the island. In the light gray regions, the ability of control depends on the initial state because of hysteretic behavior. These regions can be approached when the calculations are started in a regime of successful control.

Control Domain in (K, R) Plane   So far, in all control experiments (simulations) the FP mirror reflectivity R has been left fixed to $R = 0.76$ (0.7). The question arises, whether this value is a good choice. In order to uncover the role of R, control domains in (K, R) plane at proper φ are determined. Fig. 6.5 (c) shows the domain for fixed $\varphi = 0.74$. In order to detect hysteretic behaviour, K is varied in small steps forward and backward between zero and $K = \sqrt{R}$, using as initial value for the calculations always the final state of the previous step. Control is identified by the occurence of the period-1 pulsation in the powerspectrum, and by the simultaneous dropping of the control signal below a level of 10^{-5} of the laser emission. Bistable regions (gray) are identified by the occurence of hysteresis.

The curve $R = K^2$ (dotted line in Fig. 6.5 (d)) gives a restriction for the feedback strength $K = \sqrt{R}K_{\text{coupling}}$. The coupling strength in the latency path K_{coupling} cannot exceed $K_{\text{coupling}} = 1$ without additional amplification. The curve $R = K^2$ describes the effect of the solitary FP without attenuation in the latency path. When increasing R from zero and thus moving along the curve, control is attained at

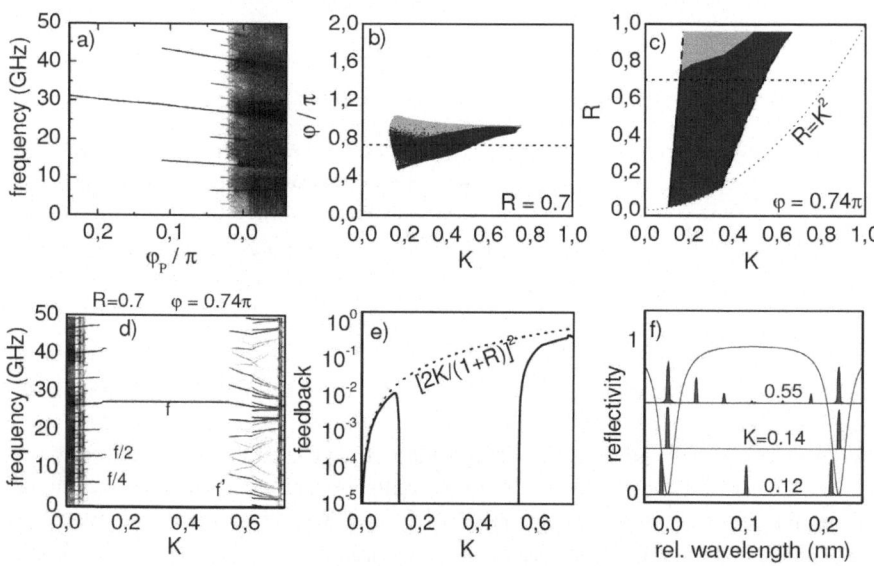

Figure 6.5: *Control of chaos in simulations. (a) Period doubling route to chaos in the ITL. Bifurcation parameter: internal phase shift φ_P in the passive section. (b), (c) Domain of control in (φ, K) and (K, R) plane for fixed R and φ, respectively. $\varphi_P = 0$. Horizontal dotted line: cut (see (d), (e)). Light gray: bistable region. (c) Domain boundaries: supercritical (solid) and subcritical (dashed) period doubling, supercritical Hopf (torus) bifurcation (dash-dotted). Dotted line: restriction for K. (d) Evolution of the powerspectra when increasing K from zero for fixed values of φ and R. (e) Feedback corresponding to (d) in units of the ITL output. Dashed curve: maximum available feedback for given K and R. (f) Optical spectrum compared to the FP reflectivity spectrum for selected values of K. The optical peaks are broadened for better visibility.*

a certain lower critical R, and later lost again. The lower threshold reflects the fact that the control force must have a certain strength to influence the dynamics. At the upper border, off-resonant spectral components become undamped by the increased feedback from the wings of the reflection minimum. By introducing an attenuating element (filter) in the latency path, K can be reduced in order to recover control.

Control Boundaries   Another question concerns the way how control is attained and lost. Fig. 6.5 (d) - (f) illustrate the control scenario along a line of given $\varphi = 0.74$ and $R = 0.7$ (dotted horizontals in Fig. 6.5 (b),(c)). When increasing K from zero, the laser undergoes a series of backward period-doubling bifurcations until the controlled period-1 state is reached at $K = 0.14$ (Fig. 6.5 (d)). Within the control range, the pulsation frequency f is locked to $1/\tau$. The corresponding reflected control power (Fig. 6.5 (e)) falls to more than five orders below the laser emission. The optical spectrum in the control domain is a comb of lines fitting exactly to the FP resonances ($K = 0.14$ in Fig. 6.5 (f)). At $K = 0.55$, control is lost in a Hopf bifurcation. Here, a torus with a second frequency f' is born. In the powerspectrum (Fig. 6.5 (d)), additional components appear at f' and its higher harmonics and mirrors. These feedback induced off-resonant components increase the reinjected control signal, and finally the torus breaks up into chaotic behavior.

An extended analysis shows that at the high-K border the domain is always limited by a supercritical Hopf bifurcation with a soft transition to quasiperiodicity. The low-K border is more complex. It is constituted by a period doubling bifurcation. Beyond a certain FP reflectivity R, this period doubling bifurcation becomes subcritical, and bistability occurs (gray region in Fig. 6.5 (c)). In this range, the stabilized period-1 pulsation coexists with feedback-modified complex dynamics. The nature of the upper border of the bistable region could not be determined here.

The role of R in conventional DFC has been analyzed in [JRRB99, vLBJ04]. Control domains in (K,R) plane have been determined, which are quite similar to the domains presented above. In general, growing R increases the range of control versus K. However, in contrast to this work, a bistability for too large R is found at the high-K border. The Hopf bifurcation confining the domain at high-K side can change from super- to subcritical type beyond a certain value of R. At the lower border, bistability is absent. This discrepancy can be either caused by a specific constellation of internal parameters in the optical setup under study, or it can present a fundamental difference between DFC of periodic orbits and DFC of modulated waves. This remains an open question here.

6.3 Summary

The original aim of this thesis - the realization of all-optical control of chaos - is achieved. An unstable SP of $1/T = 13$ GHz frequency is stabilized in an otherwise chaotic point of operation in the ITL by means of optical feedback from an external FP cavity. A period-2 SP is traced into chaos along a range of $\Delta I_P = 1.2$ mA. In chaos, the controlled period-2 SP could be increased in height by two orders of magnitude compared to the freerunning peak, and becomes much sharper. With a timescale of about $T \approx 77$ ps, this represents the fastest chaos ever been controlled experimentally. The achieved control has been found robust against slight parameter drifts, and is easily reproducable.

Like in all previous control experiments, the latency phase φ plays a significant role. Control is only possible in a finite part of a phase period. Beyond this range, the feedback can even increase the diffusion in the chaotic regime, and thus lead to less-structured powerspectra. Tuning of φ allows for a switching between either regular dynamics or even more irregular emission. A possible application of this effect could be latency switching between regular and irregular modes, e.g. in cryptography. Such a switching could be realized by a programmable piezo actor, or by refractive index modulation in the latency path.

Numerical simulations confirm the experimental success, and beyond that, show control of the period-1 SP and the period-4 SP within chaos as well. The parameter space in K, φ and R is studied comprehensively for the case of the stabilized period-1 SP in chaos. In particular, domains of control in (φ, K) and (K,R) plane are evaluated in regard to hysteresis. In all cases, for large values of R a region of bistability is found that decreases the range for control. Thus, a too large choice for R is not favourable. So far, a value of $R = 0.76$ is used in the control experiments. Actually, this value is only slightly below the values for R where bistability is observed in the calculations. Thus, the use of smaller reflectivities R in following experiments seems promising.

The computed control domains in (K,R) parameter space confirm studies on conventional DFC [JRRB99, vLBJ04]. However, a bistable region in conventional DFC always occurs at the high-K border, while bistability in optical DFC occurs at the opposite border of low-K. The cause of this discrepancy remains an open question.

7 Control of a Torsionfree Orbit

Originally, the previous Chapter was planned to be the last Chapter. In the course of the present work, all-optical DFC has been applied experimentally to quite different dynamical regimes. The known restrictions for DFC have been considered in the past experiments, especially the so-called 'odd-number limitation' [Gio91, Ush96, JBO$^+$97b, Nak97, NU98b]. The term refers to a prominent limitation of DFC: the inability to stabilize orbits without torsion. Consequently, unstable states without torsion were carefully avoided during the work on this thesis. The experiments were completed with the successful control of an unstable orbit (with torsion) in chaos.

Meanwhile, the odd-number limitation was refused theoretically [FFG$^+$07]. The authors give a counterexample using a complex feedback gain $\bar{K} = Ke^{i\varphi}$. In particular, a nonvanishing feedback phase φ is necessary. While a feedback phase is no natural parameter in conventional DFC, the all-optical implementation of DFC involves such a phase shift - the well-discussed latency phase φ. Hence, this work closes with numerical simulations and experimental investigations of all-optical DFC of a torsionfree periodic orbit.

7.1 The Odd-Number Limitation in DFC

Brief History

In 1997, two studies appeared, claiming that Pyragas control is impossible for orbits without torsion [JBO$^+$97b], or rather if the linear variational equation about the target orbit has an odd number of real characteristic multipliers greater than unity [Nak97]. Both claims refer to the same type of orbits. A generic mechanism for creating orbits of this type is a subcritical Hopf bifurcation of a stable equilibrium. Above restriction became common belief and was referred to in the following as 'odd-number limitation'. Different modifications to DFC were proposed to avoid the odd-number limitation: an oscillating feedback [SS97], a half-period delay for the special case of symmetric orbits [NU98a], and a so-called unstable DFC [Pyr01, PP06b], which was implemented successfully in electronic circuits experiments [HSC$^+$07, TMPP07].

In [Nak97], the odd-number limitation is proven for an arbitrary $n \times n$ feedback gain matrix K on the basis of Floquet theory[1], and it is claimed that for no choice of K stabilization can be achieved. However, almost all studies on DFC consider a one-dimensional problem - with a scalar K - or a multidimensional problem with a simple diagonal coupling of the control force by a diagonal matrix K. The choice of the coupling of the control force to the dynamical degrees of freedom is addressed only in few works [BS96, FBS99, BAS$^+$02], even though it presents an important aspect in control theory. This question naturally arises for spatiotemporal control setups. Further, the case of a nonvanishing coupling phase as in [RP04] and in all-optical DFC [SHW$^+$06] leads to a nondiagonal coupling.

The question of the particular choice of the feedback gain matrix K is addressed again by Fiedler in 2007 yielding a surprizing result [FFG$^+$07]. The authors provide a counterexample to the claim in [Nak97]. They demonstrate successful control of an UPO arising from a subcritical Hopf bifurcation by applying DFC with a nonvanishing coupling phase to a Hopf normal form. The circumvention of the odd-number limitation is analyzed further in [JFG$^+$07, EJ09] and is demonstrated also for subcritical Hopf bifurcations in the Lorenz equations [PS07], for rotating waves near a fold bifurcation [FYF$^+$08], for spatio-temporal dynamics [KHF$^+$09], and for a generic n-dimensional system [BPS10]. In [BPS10], the authors show that if the feedback strength exceeds a critical value there always exists a choice for the feedback phase where control is possible. The until then less regarded question of the coupling scheme appears as crucial point in the control of torsionfree orbits. An experimental verification appeared only very recently, cloning directly the equations used in [FFG$^+$07] by an electronic circuit [vLBJ10].

Topologic Illustration of the Odd-Number Limitation

Following [JBO$^+$97b], the mathematically quite complex odd-number limitation can be illustrated by topologic arguments.

As already discussed in Chapter 5 (see also Fig. 5.1), the behavior of trajectories in phase space close to an unstable orbit is governed by Floquet exponents $\lambda_k + i\omega_k$. The real part λ_k steers the instability and, thus, the expansion of the trajectory from the orbit in the respective dimension. λ_k is usually called the Lyapunov exponent. Whether the orbit is stable or unstable is determined by the largest Lyapunov exponent λ_0. If $\lambda_0 > 0$, a perturbation grows exponentially with time.

---

[1] Actually, the theorem is proven only for non-autonomous systems, and the author claims in a footnote that the proof is analogous for autonomous systems. In [JBO$^+$97b] also only non-autonomous systems are considered. For these systems, the odd-number limitation is valid. All counterexamples employ autonomous systems.

The orbit is unstable, and neighboring trajectories move away from it. If $\lambda_0 <$ 0, the orbit is stable, perturbations get damped. In case of $\lambda_0 = 0$, the orbit is marginally stable. The imaginary part ω_k describes the torsion of the trajectory, i.e. its revolution around the orbit. In one period T the trajectory rotates by an angle $\omega_k T$ about the orbit (Fig. 7.1 (a)).

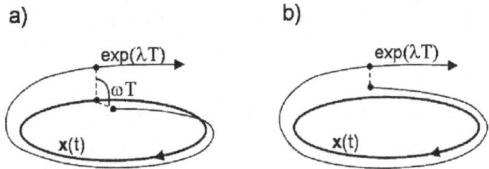

Figure 7.1: *UPO with (a) and without torsion (b). Sketch of the phase space with UPO (thick black line) and a neighboring trajectory (thin black line). The dotted line marks the distance between two points on the trajectory delayed by one period.*

When applying delayed feedback in the form (1.1), corresponding Floquet exponents $\Lambda + i\Omega$ for the orbit with feedback can be derived. In [JBO$^+$97b] and [vL10] it is argued as follows that a finite torsion is necessary for successful stabilization by DFC. The Floquet exponent determines the distance between two points on a neighboring trajectory delayed by one period T (see black dots in Fig. 7.1). This distance corresponds to the time-delayed difference which builds the control force F in DFC. Thus, the magnitude of the control force is directly proportional to this distance. When the UPO with time-delayed feedback is about to become stable, the Lyapunov exponent has to cross zero: $\Lambda = 0$. The orbit is about to become attractive, but the trajectory has not reached it yet. If the torsion is also zero, the two points in Fig. 7.1 coincide and the control force disappears. But the orbit is not stable yet, it is only marginally stable. This way, the trajectory cannot reach the orbit, and control fails. Simplified, this is the essence of the odd-number limitation in DFC.

Circumvention of the Odd-Number Limitation

The circumvention of the odd-number limitation has been demonstrated with the normal form of a subcritical Hopf bifurcation [FFG$^+$07, JFG$^+$07, EJ09], for subcritical Hopf bifurcations in the Lorenz equations [PS07], for rotating waves near a fold bifurcation [FYF$^+$08], for spatio-temporal dynamics [KHF$^+$09], and for a generic n-dimensional system [BPS10]. In the following, the involved stabilization mechanism is described.

The torsionfree UPO emerging from a subcritical Hopf bifurcation can be stabilized by DFC if the control force is coupled by a nondiagonal control matrix $Ke^{i\varphi}$ with a nonvanishing feedback phase φ [FFG$^+$07, JFG$^+$07]. Stabilization fails for zero phase φ. Besides the crucial role of the feedback phase, the dependence of the oscillation frequency on the amplitude, parametrized by γ [JFG$^+$07], plays an important role. Stabilization requires a negative drift of the oscillation frequency with increasing amplitude. A substantial dispersion, i.e. a large value of $|\gamma|$ is favorable. The two requirements of a substantial dispersion γ and of a nonvanishing feedback phase φ can be illustrated topologically as well [vL10]. Again, the point $\Lambda = 0$ is considered. When the system shows a substantial dispersion, then two points on a neighboring trajectory delayed by one control period τ do not coincide, but have a finite distance. Thus, always a finite control force remains, but this force is directed along the trajectory. By additionally using a nondiagonal coupling scheme with a finite coupling phase φ, the direction of this remaining control force F can be changed such that F contains a part directed towards the orbit. Stabilization of the UPO is possible.

In [FFG$^+$07, JFG$^+$07], the bifurcations involved in the stabilization process are revealed. The odd-number limitation is broken by the occurence of a transcritical bifurcation where the torsion-free UPO collides with another delay-induced periodic orbit and both orbits exchange stability. In the subcritital Hopf bifurcation, a stable stationary state loses stability, and an UPO is born. When adding delayed feedback, another Hopf bifurcation of the bifurcating stationary state appears. Here, the stationary state becomes unstable and a stable periodic orbit arises. This orbit is induced by the additional delay, and for it the feedback does not vanish. When the feedback strength exceeds a critical value, the UPO and the delay-induced orbit collide in a transcritical bifurcation and exchange stability. Surprizingly, the same sequence of bifurcations is observed in the higher-dimensional Lorenz example [PS07]. The stabilization mechanism is further generalized in [BPS10].

The optical experiments presented in the following exceed previous work in several respects. So far, extended DFC with multiple delays as in the present setup has not been studied in this context. Furthermore, in the studies [FFG$^+$07, JFG$^+$07, PS07, FYF$^+$08, EJ09, BPS10, vLBJ10] a control force tangent to the two-dimensional center manifold of the Hopf bifurcation is considered. A global control force is used in [KHF$^+$09]. In contrast, in all-optical DFC a control force proportional to the electric vector of the optical field, i.e. purely optical, is used to stabilize relaxation oscillations (RO) which involve carrier densities and photon densities. This control force is neither global nor tangent to the center manifold of the bifurcation.

7.2 Subcritical Hopf Bifurcation in the Freerunning Laser

In a subcritical Hopf bifurcation, a stationary state (focus) becomes unstable and an unstable periodic orbit (UPO) is born (black lines in Fig. 7.2 (c)). Beyond the bifurcation, the dynamics typically show a very abrupt switch to a stable large-amplitude limit cycle (LC) nearby in phase space. The UPO born in this bifurcation diverges backwards and can, e.g., annihilate with a stable limit cycle in a saddle-node (SN) bifurcation. This UPO has one single real-positive Floquet multiplier, thus being not stabilizable by standard DFC. Usually, the switch from steady state to the limit cycle and back shows hysteresis.

In what follows, a subcritical Hopf bifurcation due to undamping of relaxation oscillations (RO) is studied in the freerunning amplified feedback laser (AFL). This type of bifurcation has been predicted by the bifurcation analysis of device-realistic models of lasers with short-delay feedback [Sie02, SRS, BBK+04]. Experimentally, the appearance of subcritical Hopf bifurcations has already been verified in the AFL [BBK+04, UBB+04]. Here, a sudden onset of large-amplitude selfpulsations (SP) was observed, but no hysteresis. The device studied in the following also shows such a transition.

Fig. 7.2 (a) and (b) picture the measured bifurcation scenario. Bifurcation parameter is the current I_P in the passive section. Initially, the laser is operated in cw emission. Here, the omnipresent internal noise excites deviations from the stationary state. The laser returns to equilibrium via damped RO, which are manifested in the powerspectrum by a weak peak around 8 GHz (lowest panel in Fig. 7.2 (a)). Oscillation amplitude and damping of the RO are represented by area and width of this peak, respectively. The linear decrease of the damping Δf (diamonds in Fig. 7.2 (b)) shows that this transition is indeed a Hopf bifurcation. The damping passes zero at the critical current I_P^H which is obtained by linear extrapolation of the measured data. While approaching the bifurcation, the amplitude of the damped RO (black dots) increases slightly and suddenly jumps to a high level when the laser switches to stable large-amplitude SP of about 7 GHz frequency.

This sudden jump is fingerprint of a subcritical bifurcation. However, no hysteresis is found on tuning I_P backwards. The case of a degenerate Hopf can be excluded, because the jump occurs already before the bifurcation.

Fig. 7.2 (a) gives an expanded view on the evolution of the measured power-spectra around the jump. The transition from RO (bottom) to SP (top) occurs continuously. From bottom to top, the SP peak at 6.8 GHz grows on expense of the RO peak at 7.9 GHz. In the mid panels, the two states appear simultaneously in pow-

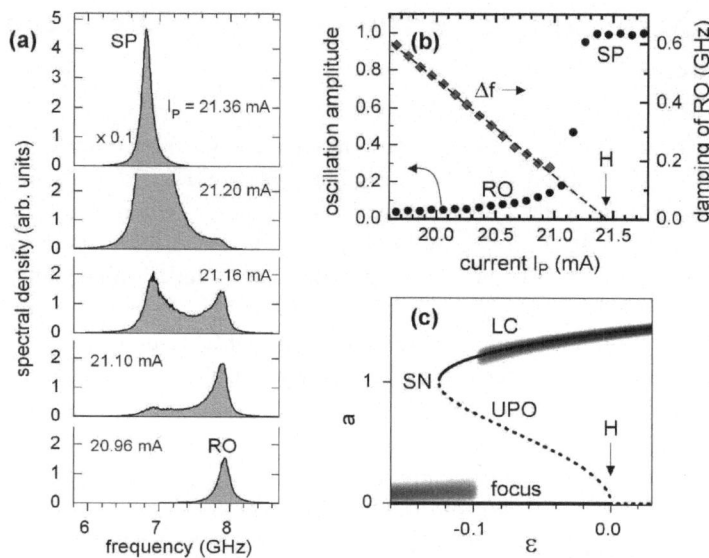

Figure 7.2: *Subcritical Hopf bifurcation in the AFL. (a) Powerspectra of the freerunning AFL for selected currents I_P. (b) Dots: amplitudes of RO and SP normalized to maximum vs. I_P. Diamonds: damping of RO; the position of the Hopf bifurcation H is obtained by linear extrapolation to zero. $(I_L, I_A) = (87.1, 8.7)$ mA. (c) Corresponding scenario of a Duffing-van der Pol oscillator [ZVA⁺10]. Lines: normalized amplitudes of deterministic oscillations and focus $(a = 0)$ versus bifurcation parameter ε. UPO: unstable periodic orbit. SN: saddle-node bifurcation of orbits. LC: stable limit cycle. Gray (red) scale coded: normalized amplitude probability distribution (7.1) for noise level $D = 10^{-3}$. Probabilities below 1% are not shown.*

erspectrum. These RO spectra are measured with a sampling time of 3 seconds. This sampling time exceeds the timescale of the intrinsic laser dynamics by more than 8 orders of magnitude. Thus, it is supposed, that the two peaks originate from alternating epochs of RO and SP operation occurring during the sampling time of the spectrum analyzer. Above spectra are just linear superpositions of pure RO spectra and pure SP spectra with different relative weights. The peak area reflects the mean residence time in the respective state. The two residence times strongly change with the current I_P in opposite directions. They are comparable to each other only in the region between $I_P = 21.1$ mA and $I_P = 21.2$ mA.

These findings can be explained by a noise-induced switching between the two states, RO and SP, which coexist in the underlying deterministic Hopf bifurcation.

A similar result has been found in recent literature [ZVA⁺10]. The authors study a subcritical Hopf bifurcation in the stochastic Duffing-Van der Pol oscillator with linear damping ε (Fig. 7.2 (c)). In the Hopf bifurcation H at $\varepsilon = 0$ a deterministic UPO (dotted line) is born. The UPO bends back and annihilates with a stable limit cycle (LC) in a saddle-node (SN) bifurcation at $\varepsilon = -0.125$. The amplitudes a of the two branches are given analytically by $a = \sqrt{1 \pm \sqrt{1 + 8\varepsilon}}$. With additional weak noise, the bistability of focus and LC is destroyed, because noise-induced transitions between the two states occur from time to time. The probability distribution of oscillation amplitudes can be approximated analytically [ZVA⁺10]

$$p(a) \sim a\exp\left(-\frac{a^2(a^4 - 3a^2 - 24\varepsilon)}{48D}\right). \tag{7.1}$$

Here, the same qualitative behavior as in above AFL measurements is observed. Except for a tiny transition region with equal probability in both states (here close to $\varepsilon = -0.1$), the oscillator spends nearly all time either in a noisy LC or near the focus. Thus, the observed disappearance of hysteresis is just phenotype of a subcritical Hopf bifurcation when substantial noise is present.

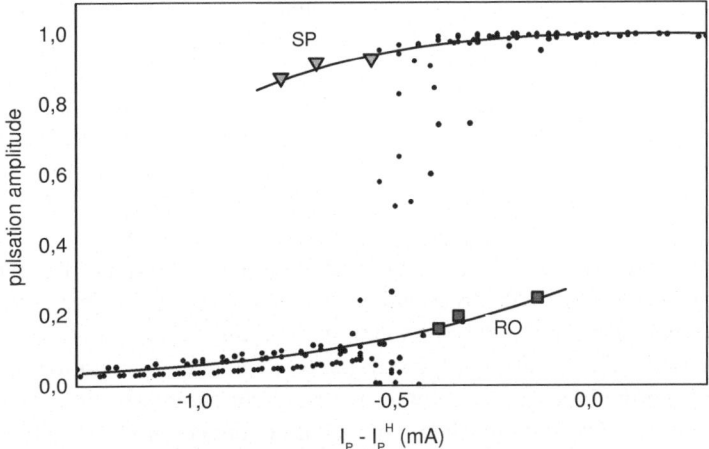

Figure 7.3: *Noninvasive control of bistability at the subcritical Hopf bifurcation. Focus and SP branch are extended by all-optical DFC. The normalized pulsation power versus distance to bifurcation is summarized for several points of operation. Black dots: freerunning laser. Gray triangles: SP stabilized with a 6.9 GHz etalon. Gray squares: focus stabilized with a 17 GHz etalon. $R = 0.5$, $K = 0.02$.*

7.3 Recovering Hysteresis by Optical Control

In what follows, noninvasive optical DFC is applied to prove that the Hopf bifurcation discussed above indeed possesses a bistable region. The developed experimental tool for all-optical DFC presents a means for the noninvasive stabilization of genuine laser states, i.e. states which exist also in the freerunning system. In the following, it is used to stabilize the missing branches of the hysteresis loop against noise-induced escapes.

Control of SP against Noise

The laser is first operated in a regime of stable cw emission slightly left from the amplitude jump. Here, the noise-destabilized SP shall be recovered by optical DFC. The pulsation frequency of the stable SP right from the nearby jump is 6.9 GHz. A FP etalon with matching roundtrip frequency $1/\tau = 6.9$ GHz is added to the cw emitting laser. The latency phase φ is varied until a regime is found where the laser shows SP operation and the FP transmission is maximal. The optical frequency and the pulsation frequency are precisely matched to the FP resonances by a fine-tuning of I_A and the laser temperature T. This way, the SP could indeed be stabilized previous to the jump, accompanied by maximal transmission behind the FP.

The stabilized SP is traced along the missing hysteresis branch as follows. Stabilization of the SP against cw emission is repeated in two different distances from the jump using the same FP etalon. Because the FP parameters are fixed, but optical frequency and pulsation frequency shift with I_P, this procedure requires slight modifications of the point of operation (via I_A and temperature). After each control measurement, the complete bifurcation sequence without FP is measured. Fig. 7.3 pools the data for the freerunning bifurcation sequences (small black dots). For better comparison, the pulsation amplitude in this figure is normalized to the maximal value of a series, and the bifurcation parameter I_P is normalized to the distance from the bifurcation. Obviously, the slight parameter variations did not alter the regime of operation. The SP branch is continued into the regime of stable cw emission by the three control experiments (gray triangles). Since the achieved stabilization is noninvasive, a corresponding SP must exist in the phase space of the underlying deterministic system without feedback. Probably, the underlying deterministic SP could have become unstable on the way across the measured amplitude jump due to another bifurcation. However, as there is no indication for other bifurcations, it is concluded that the deterministic SP remains stable and the jump is due to the noise-induced escape described above.

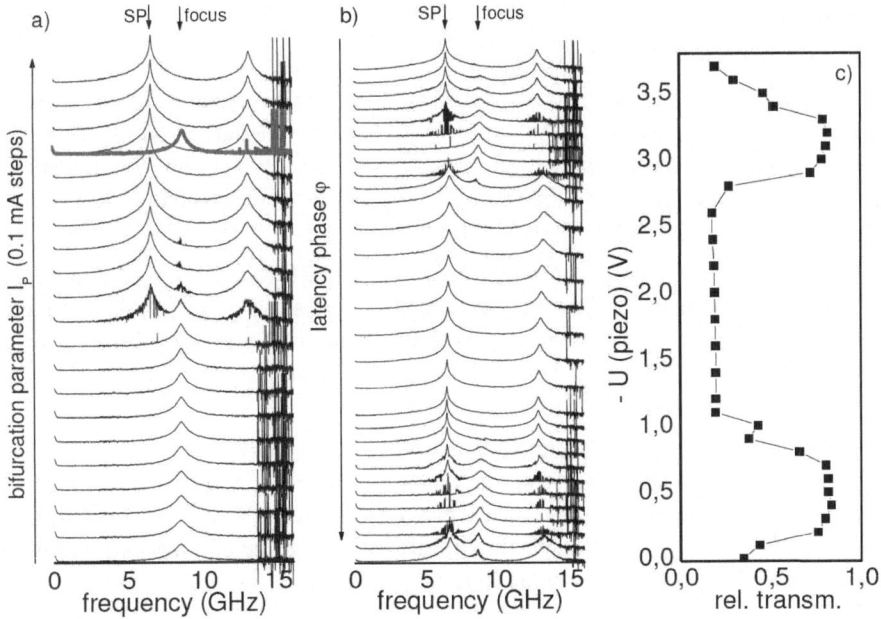

Figure 7.4: *Control of focus against noise. (a) Black powerspectra: freerunning bifurcation versus I_P, which increases in vertical direction. Thick gray line: powerspectrum of the controlled focus after the jump. (b) Powerspectra under control versus φ. (c) Relative transmission corresponding to (b). All spectra plotted in logarithmic scale.*

Control of Focus Against Noise

The hysteresis is completed right from the jump in the regime of stable SP. In the freerunning bifurcation sequence, the coexistence of SP and cw emission is indicated by spikes around the RO frequency closely after the jump (black powerspectra in Fig. 7.4 (a)). The noise-destabilized focus beyond the jump shall be recovered by optical control.

Stabilization of unstable stationary states by DFC has been discussed in detail in Chapter 4. In the case of stationary target states, the FP delay τ must not equal the RO period T of the focus, but should optimally be half this value [HS05]. The RO frequency of the focus is approximately $1/T = 8$ GHz. Thus, a FP etalon with $1/\tau = 17$ GHz roundtrip frequency is added to the laser, which is operated in selfpulsating mode in 0.7 mA distance from the jump (see the fourth spectrum from top in Fig. 7.4 (a)). Indeed, after resonant adjustment, the laser with FP shows cw emission, accompanied by maximal transmission behind the FP (thick gray powerspectrum in Fig. 7.4 (a)).

Repetition of the experiment with the same FP in two different distances from the jump continues the hysteresis branch. Slight accommodations of the laser parameters are necessary to maintain resonance with the FP. The three control measurements are summarized in Fig. 7.3 (gray squares), together with the corresponding freerunning bifurcation sequences (dots).

A variation of φ, starting in a regime of a stabilized focus, leads to a periodic exchange between the stabilized focus and SP operation (Fig. 7.4 (b)). In the regions of controlled stationary state, the FP transmission is high (Fig. 7.4 (c)), whereas in the SP regions the laser receives strong feedback from the FP. φ can be used to switch between these two states here.

Above control experiments uncover the coexistence of cw and SP emission before the Hopf bifurcation. The subcritical nature of the Hopf bifurcation can therefore be considered as proven. Thus, there must branch off an UPO with an odd number of positive Floquet exponents. In the following, the noninvasive stabilization of this UPO by optical DFC is studied.

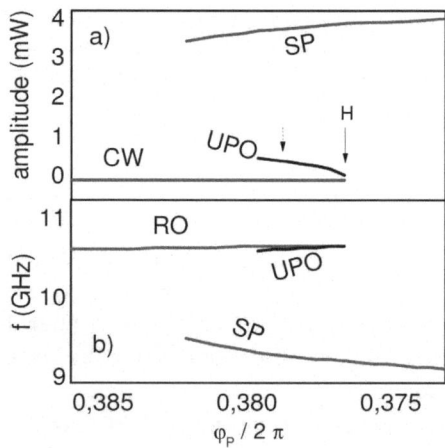

Figure 7.5: *Simulations: All-optical control of UPO at a subcritical Hopf bifurcation. (a), (b): Optical power range and RO frequency versus the internal phase shift φ_P in the AFL. Gray: freerunning laser. Black: control.*

7.4 Control of the Torsionfree UPO in Simulations

The planned experimental control of the torsionfree UPO faces more difficulties than all previous control experiments. Pulsation period and optical frequency of the UPO in noticeable distance to the bifurcation are unknown. Moreover, if at all, control is only possible in a extremely limited range of control parameters [FFG+07], which has to be identified first for this system. Thus, control is initially investigated in numerics.

Bifurcation Scenario in Simulations

Fig. 7.5 shows a subcritical Hopf bifurcation in the freerunning simulated AFL. Bifurcation parameter is the internal phase shift φ_P (gray lines). φ_P is changed in small steps ($10^{-3}\pi$), and the attractor reached in one step serves as initial state for the next step. By changing φ_P up and down, the laser without noise shows the expected hysteretic behaviour. In a bistable region, both cw emission and the large-amplitude SP exist. The RO frequencies in Fig. 7.5 (b) are determined from the transients towards cw emission. Please note that here the SP frequency is below the RO frequency. This is a prerequisite for circumventing the odd-number limitation.

Control of UPO

The torsionfree UPO itself cannot be observed directly in the freerunning simu-lated AFL. It is expected to connect the focus and the SP branch. The laser is first set to cw emission only slightly left from the bifurcation point. Control is applied by adding two passive waveguide sections to the AFL structure, realizing the FP and the latency roundtrip. Because RO frequency and wavelength of the UPO are unknown, the FP parameters are first chosen resonantly to the weakly damped RO of the cw state. Close to the bifurcation point, this presents the best guess for the unknown UPO. The challenge is to find a control parameter combination of K, φ, and R that undamps the RO of the UPO. In order to avoid narrow FP resonances, and thus unwanted feedback-induced disturbances, a rather low value of $R = 0.3$ is chosen. The feedback strength K and the latency phase φ are varied until the RO become undamped and settle to a stable small-amplitude oscillation. By it-erative adaptations of the FP parameters, the feedback power is decreased until resonance with the FP is achieved. The procedure is stopped when the feedback signal reaches a level below a 10^{-5}-fraction of the 10-mW AFL output, which is surely not affecting the internal laser dynamics. The thus found values for K and

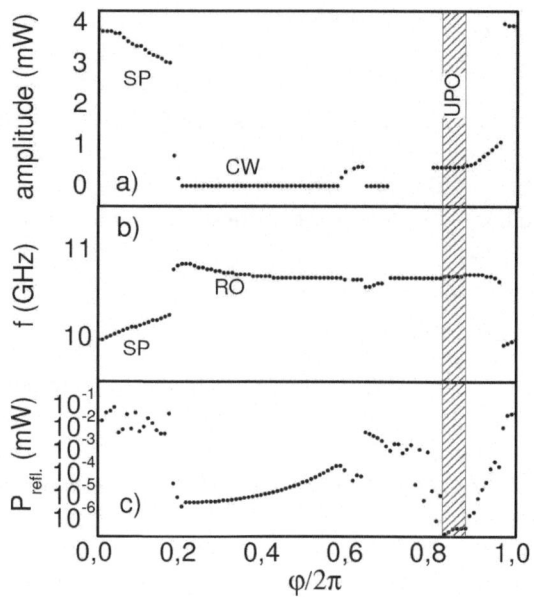

Figure 7.6: *Simulations: Control of torsionfree UPO versus phase. (a)-(c): Power range, RO frequency, and power reflected by FP versus the latency phase* φ*. Gray: freerunning laser. Black: control. Shaded area: noninvasive control of UPO. The bifurcation parameter* φ_P *is fixed to 0.3785.*

φ are held fixed in the following. The UPO is traced backwards by decreasing the bifurcation parameter φ_P in small steps. The FP parameters are readjusted in each step to maintain resonance. This procedure yields the black lines in Fig. 7.5. Beyond a certain distance from the bifurcation, control is lost due to the increased instability of the UPO.

Domain of Control

In a next step, the domain of control versus φ is explored. Fig. 7.6 shows the effect of a variation of φ in the point of operation marked by a dashed arrow in Fig. 7.5. Increasing $\varphi/2\pi$ from zero, various states appear, including large regions of SP operation as well as cw emission. At 0.8 (the hatched window in Fig. 7.6), a small amplitude oscillation exists, accompanied by negligible feedback power. This small-amplitude oscillation represents the noninvasively stabilized UPO. At the left border of the window, control is lost due to the onset of a very slow amplitude oscillation. Here, no amplitude is drawn. At the right border of the window, the small-amplitude oscillation is no longer resonant with the FP, and the amplitude of the undamped RO starts to increase. This feature indicates the transcritical bifurcation which breaks the odd-number limitation [FFG+07, JFG+07, FYF+08, KHF+09, EJ09, BPS10].

The control range of the UPO found in these simulations is very small, but this meets the theoretical expectations [FFG+07]. The control range versus φ covers only a five per cent fraction of one period. Moreover, further computations show that the control gain is restricted to $0.05 < K < 0.07$. In above calculations, the latency time τ_l is set to zero, but fortunately control could be also attained with $\tau_l = \tau$. Control is also tested with a larger reflectivity $R = 0.5$ (the next value available in experiment), but here no stabilization of the UPO is achieved.

In above simulations, parameters similar to table 2.3 are used, except for:
$[L_A, L_P, L_L] = [220, 500, 220]\ \mu\text{m}$, $\kappa = [0, 0, 1] x (250 + 6i)\ \text{cm}^{-1}$,
$\gamma = [25, 25, 25]\ \text{cm}^{-1}$, $\delta = [22147, variable, 19126]$, $I = [9, 0, 80]\ \text{mA}$,
$\varepsilon = [1, 0, 1] 10^{-24}\ \text{m}^3$.

7.5 Experimental Control of a Torsionfree Orbit

The presented numerical experiment demonstrates the ability of the all-optical setup to control torsionfree UPOs, as predicted in [FFG+07]. However, it also points out the challenges which have to be met in experiment. The domain of control is extremely small, requiring a very accurate setting of the relevant parameters. Above calculations on the one hand revealed how subtly the control parameters have to be met, but on the other hand enable simulation aided experiments. The simulations showed that control is expected only for a few-percent range of φ. Thus, the setup is improved by introduction of a new position-stabilized piezo actor with higher resolution (20 nm). The calculations revealed the location of the UPO versus φ: with increasing φ it is expected after a stable focus region, and just before a jump to high amplitude SP. This phase tuning scenario will be used as a kind of fingerprint to find the UPO.

The AFL is operated close to a subcritical Hopf bifurcation in a regime of damped RO with 7.9 GHz frequency. The RO frequency of the UPO is expected to be slightly smaller than the damped RO of the cw state. Thus, a FP etalon with 7.7 GHz round-trip frequency is selected. The FP mirror reflectivity R is chosen equal to the value $R = 0.3$ used in numerics. The latency time τ_l is set equal to the FP delay. The feedback strength K cannot be derived directly from the experimental setup with the required accuracy. While the setup allows for a sensitive tuning

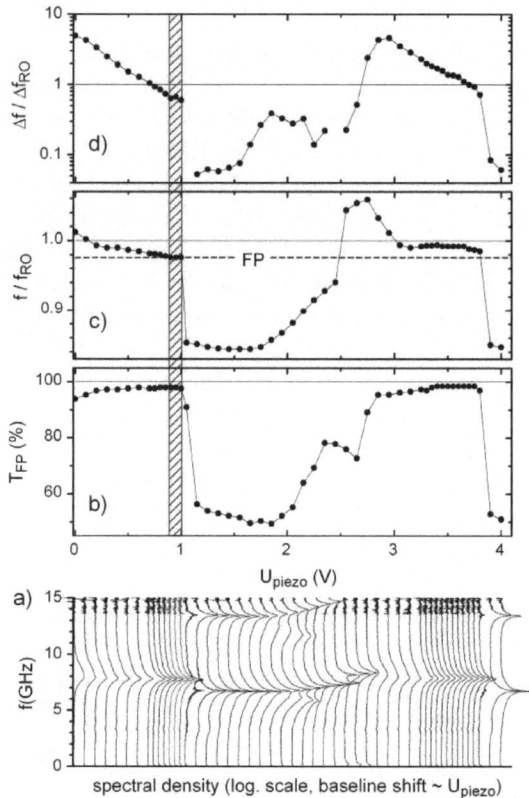

Figure 7.7: *Control of UPO in dependence on* φ. *(a) Powerspectra (logarithmic scale) versus latency phase* φ *(in lateral direction). (b)-(d) Relative FP transmission, RO frequency, and RO peak width, respectively, versus piezo voltage (black dots). Dotted lines: freerunning laser. Solid line: FP frequency. Shaded area: noninvasive control of UPO.*

of K by a variable filter, the absolute value of K is also affected by the lens and, strictly speaking, stays uncertain. In order to correlate the feedback strength K with the filter attenuation K_{filter}, following overall measurements and simulations are carried out first. The numerical experiment shown in Fig. 7.6 is repeated for various values of K between $K = 0.01$ and $K = 0.2$. For each value of K, power spectra of the emitted output in dependence on φ are calculated. Then, a similar measurement is performed: the FP is coupled to the AFL in cw emission just left from the amplitude jump. The emission wavelength of the cw emitting laser is

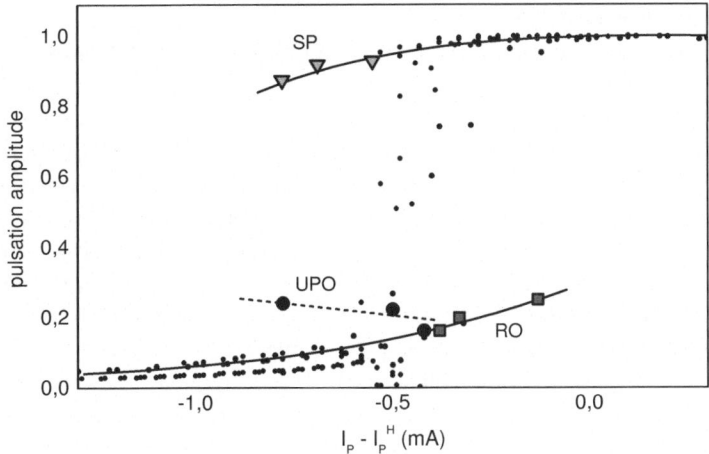

Figure 7.8: *Control experiments at the subcritical Hopf bifurcation. (a), (b) Normalized pulsation power versus distance to bifurcation summarized for several points of operation. Open symbols: freerunning laser. (a) Noninvasive control of bistability. Focus and SP branch are extended by all-optical DFC. Black squares: SP stabilized with a 6.9 GHz etalon. Black triangles: focus stabilized with a 17 GHz etalon. (b) Stabilization of UPO (black dots) with a 7.7 GHz FP etalon. (c) Power spectra of freerunning laser (filled gray area) and controlled UPO (black line) at point A in panel (b). (d) Frequency deviation between UPO and focus vs distance from bifurcation. (e) Peak width of UPO (black dots) and focus (gray squares) vs distance from bifurcation.*

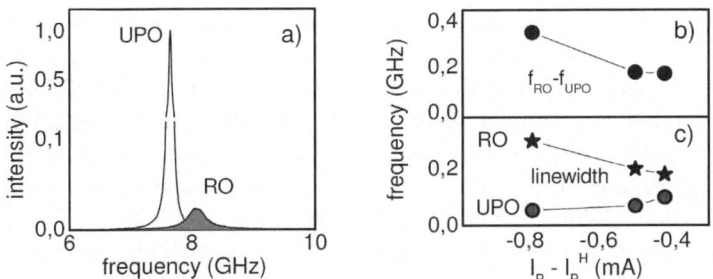

Figure 7.9: *(a) Power spectra of freerunning laser (filled gray area) and controlled UPO (black line). (b) Frequency deviation between UPO and focus vs distance from bifurcation. (c) Peak width of UPO (black dots) and focus (open squares) vs distance from bifurcation.*

adjusted to resonance. Then, for various values of the filter attenuation K_{filter}, φ is tuned, and power spectra of the emitted output are detected. Comparison of the measured series with the simulated data enables an identification of the range for K_{filter} where control is expected.

After adjusting K in the above described procedure, the laser parameters I_A, T and I_P are varied, and it is searched for a similar scenario versus φ as in Fig. 7.6. This laborious approach finally leads to a point of noninvasive control. Fig. 7.7 (a) shows a series of RF spectra versus φ measured in 0.42 mA distance from the bifurcation. The series starts left in a regime of a stable focus. The spectrum shows a broad RO peak around 8 GHz, and the FP transmission (b) is already high. On increasing φ, the peak grows and sharpens, while the RO frequency drops and the transmission increases. In the hatched region, RO frequency and the peak width stay constant at 7.7 GHz and 0.1 GHz, respectively, accompanied by maximal FP transmission (Fig. 7.7 (b)-(d)). Going further, a sudden jump to high amplitude SP (two-peak spectra in Fig. 7.7 (a)) occurs while the transmission falls. The experimental behaviour is in accordance with the simulations in Fig. 7.6, where stabilization of the UPO occurs just before the onset of SP.

Various facts provide evidence that the stabilized small-amplitude SP indeed represents the target UPO. First, the frequency of the stabilized pulsation is below that of the damped RO in the freerunning cw emitting laser, but it equals the free spectral range of the FP etalon. Second, the RF line narrows significantly while approaching the control region. Third, the FP transmission is maximum ($T_{FP} = 0.98$). The associated relative control power is below $(1 - T_{FP})K^2 = 0.72 \cdot 10^{-7}$. This value is sufficiently small to conclude that the laser under control is in a state that exist also in the freerunning device (compare the discussion in Chapter 6, and Fig. 6.4). For these reasons, the RF feature in the shaded region represents an unstable small-amplitude RO of the laser, and thus the target UPO. A second control window is not found when tuning φ further. Probably, small parameter drifts have deteriorated the FP adjustment during the measurement time.

The stabilization of the UPO could be repeated in two different distances from the bifurcation, yielding the negative-slope branch in Fig. 7.8 (black circles). The power of the stabilized small-amplitude pulsation grows with increasing distance to the bifurcation. The frequency separation $f_{focus} - f_{UPO}$ from the RO peak grows as well (Fig. 7.9(b)), whereas the line width decreases (c). Fig. 7.9 (a) demonstrates how distinctly different the powerspectra of the controlled and of the freerunning laser become at the point beyond which control can no longer be achieved (about 0.78 mA before the bifurcation). The small peak showing the damped RO of the freerunning focus (gray filled) is transformed with FP to a sharp pulsation with smaller frequency (black line). Going closer to the bifurcation, the rf features of UPO and noise-driven RO become more similar until a distinction is impossible.

7.6 Summary

After the proof-of-concept experiments presented in the previous Chapters, noninvasive optical DFC is used to investigate a subcritical Hopf bifurcation in the AFL. First, cw operation as well as SP, close to a subcritical Hopf bifurcation with missing hysteresis, are stabilized against noise-induced escape. This way, the otherwise hidden bistability between both states before the bifurcation is uncovered. The latency phase φ can be used to switch between the two states, an aspect which could be interesting in communication technologies. Second, a torsion-free UPO at this bifurcation is stabilized noninvasively. This experiment confirms the invalidity of the so called odd-number limitation of DFC, extending the theoretical conclusions drawn for the model cases in [FFG+07, JFG+07, PS07, FYF+08, KHF+09, EJ09, BPS10] and the experiments with electronic circuits [vLBJ10] to practical optical systems with a much more complex phase space and a significant impact of noise. Furthermore, the results show that the control also works for a control force which is non-global and not tangent to the two-dimensional manifold of the bifurcation as assumed previously. In these experiments, a significant latency time τ_l of the order of τ cannot be avoided. However, control of a torsionfree orbit also works under these conditions, a point which has not been studied so far.

The last experiment reveals the limits of the developed experimental control setup. Much efforts were necessary to achieve the presented results. The found stabilization effect is apparent, but still weak. Comparing the experimental efforts with the previous control experiments, I conclude that the impact of the odd-number limitation is nevertheless still strong here. Although the odd-number limitation is invalid, the control succeeds only in a very small range of parameters, limiting robustness and making daily-use applications difficult.

These results are published in [SWH11].

8 Conclusion

The aim of the present thesis was the experimental study of all-optical chaos control. A control setup proposed 16 years ago [SSG94] is implemented for the first time: experimental delayed feedback control operating solely in the optical domain. Delayed feedback control [Pyr92] is a powerful method, which enables stabilization of unstable states in dynamical systems, e.g. within chaos. It is strikingly simple to realize - an output signal is compared with a temporally delayed version of itself, and fed back as input to the controlled system. Further, the control does not perturb the stabilized states - it acts 'noninvasive' in the sense that the input power is zero under control. Delayed feedback control has been applied successfully not only in physics, but in quite different fields of science. The experimental applications mainly involve electronic feedback loops, which are practical, but also restricted in speed to the sub-GHz range.

This work presents the first all-optical implementation of delayed feedback control, and at the same time the fastest realization so far. Ultrafast dynamics in a multisection semiconductor laser on the scale of tens of GHz are noninvasively stabilized by direct optical feedback from a Fabry-Perot interferometer. By these means, several proof-of-concept experiments are performed: the all-optical stabilization of an unstable stationary state, and the stabilization of unstable periodic orbits, both within chaos and outside of chaos. These results are published in [SHW+06] and [SWH08].

Beyond the proof-of-concept, new aspects appear and are studied in detail. The all-optical control scheme features an additional control parameter, compared to standard delayed feedback control: the optical phase of the feedback. The role of this new parameter is twofold: it dramatically increases the experimental efforts, but it also presents a new degree of freedom. A tunable optical phase shift allows for an easy switching of the sign of the feedback, and thus gives the possibility to change between oppositional effects like stabilization and destabilization. From the theoretical point of view, this feedback phase attracted attention with a Letter by Fiedler and coworkers [FFG+07], which was published during the work on this thesis. The authors refuse an until then believed theorem, which states that delayed feedback control is not able to stabilize a certain type of states - torsionfree orbits [Nak97]. Actually, this presents a serious constraint to the control method. According to [FFG+07], a nonvanishing phase of the feedback can eliminate this so-called 'odd-number limitation' of delayed feedback control. Consequently, the refusal in [FFG+07] is investigated experimentally. In fact, control of a torsionfree orbit, contradictory to the odd-number limitation, is achieved. But the effect is weak, and the efforts are extremely high, so the experiment rather demonstrates the still strong impact of the odd-number limitation. The results are published in [SWH11].

Some interesting further aspects, all in conjunction with noise, could be only touched shortly. First, delayed feedback control is also able to control the coherence of oscillations. Coherence control with the realized optical setup is tested successfully, but is not evaluated quantitavily. All-optical control can further influence the properties of noise-driven oscillations. The optical phase appears as a quite handy control parameter in this context. Last, all-optical control is applied to stabilize noise-destabilized states. Here, the optical feedback allows for a switching between the two states at a noise-destroyed bistable region.

All in all, this work closes a gap by demonstrating a still outstanding application of an otherwise widely stressed control method. The method is interesting from a theoretical point of view, as it enables the experimental study of laser dynamics. Possible technological applications would have to cope with the extremely high stability demands which have to be made to the setup. However, the control of chaos has been found quite robust against parameter drifts and could be easily reproduced. But, the used feedback strengths of a few per cent only enabled a tracing of the stabilized state in a current range of about 1 mA beyond the bifurcation. The method seems to fail for orbits with higher instability, or at least a larger feedback strength is required here. The fixed delay time of the feedback, caused by the use of solid etalons as Fabry-Perot cavities, limits the range of application of the method. Even a tunable Fabry-Perot interferometer with two independent mirrors cannot solve this problem satisfactorily. Commercial devices only enable a tuning of the delay by some free spectral ranges. Thus, the all-optical implementation of delayed feedback control is limited by a more or less fixed delay time. On the other hand, the fixed delay time could be of advantage in telecommunication, where a high stability of the pulsation frequency to a given value is desired.

Appendix

A Constraints of Delayed Feedback Control

Although the theoretical understanding of DFC is far from being complete, important knowledge on the mechanism and the limits of DFC have been gained in the last years. This appendix summarizes the main constraints of DFC.

A.1 The Main Control Parameters

Simple conventional DFC [Pyr92] is governed by three parameters: the feedback strength K, the delay τ, and the degree of instability of the desired unstable orbit, given by its largest Lyapunov exponent λ. The latter quantity measures how fast the freerunning system deviates from the UPO. The first limitation of the control range refers to the feedback strength K. DFC is only effective within a finite interval of K, limited by a minimal and a maximal critical feedback strength [PT93].

Highly Unstable and Long-Period Orbits Furthermore, DFC cannot stabilize arbitrary UPO's. Control fails when the product $\lambda\tau$ exceeds a certain critical value, which is estimated by Just in [Sch99] to $\lambda\tau < 2$. Orbits with long periods, or highly unstable orbits are not suited for DFC. In order to ease this constraint, advanced control strategies have been developed, e.g. the extended delayed feedback control [SSG94]. Here, the control signal consists of an infinite series of integer multiples of τ with geometrically decreasing relative weigths R^n, $0 < R \leq 1$. This way, the operation range of DFC can be extended to $\lambda\tau < 2(1+R)/(1-R)$. The memory parameter R adjusts the impact of states further in the past. In all-optical DFC, R corresponds to the mirror power reflectivity of the cavity. While in conventional DFC a growing R increases the range of control [vLBJ04], in all-optical DFC a too large R leads to a decrease of the control range (Chapter 6).

Control Loop Latency In real experimental DFC setups, a finite latency time τ_l is unavoidable. In general, a nonvanishing latency τ_l decreases the stabilizing effect [JRRB99, HS03], therefore τ_l has to be kept as small as possible. In particular, τ_l must be smaller than $\tau_{l,\text{crit}} = 1/\lambda - \tau/2$ [HS03] for successful control. Keeping the latency time small turns out as a major challenge in optical DFC. In the optical experiments, always a latency time equal to the delay time is used: $\tau_l = \tau$, except for the experiment in Chapter 4 where the latency is $\tau_l = 3\tau$.

A.2 The Odd-Number Limitation

Another constraint of DFC refers to the type of unstable orbits which can be stabilized by DFC. UPO's can be characterized by the evolution of neighboring trajectories in phase space - in simple words, by the way the system leaves the orbit. For a system with n dimensional equations of motion this is condensed in n characteristic multipliers with n so-called 'Floquet exponents' $\lambda + i\omega$. Here, the Lyapunov exponent λ measures the radial expansion from the unstable orbit. The imaginary part ω, the frequency component, gives the revolution of the trajectory around the unstable orbit. It has been shown [JBO⁺97b] that a finite frequency ω corresponding to a finite torsion of the orbit is a necessary constraint for stabilization. Formally, for orbits with an odd number of positive real Floquet exponents conventional DFC fails [Gio91, Ush96, Nak97]. This restriction is known as 'odd number limitation'. On the contrary, for orbits with maximal torsion $\omega = \pi/T$ DFC works best [JBO⁺97b]. A generic mechanism for creating torsionfree orbits is a subcritical Hopf bifurcation of a stable equilibrium. Such bifurcations exist in different systems like the Lorenz equations [PS07], or the Hodgkin-Huxley model of action potentials in neurons [GW00]. They were observed e.g. in NMR lasers [BHL89], and multisection semiconductor lasers (MSL) [BBK⁺04, UWH⁺05].

Much effort has been invested to develop modified delayed feedback schemes which avoid the odd-number limitation. The most famous and elaborated scheme is the so-called unstable controller invented by Pyragas [Pyr01, PPB04]. Other modifications are the implementation of a time-periodic feedback gain K [SS97], a half-period delay for the special case of symmetric orbits [NU98a], and, very recently, a promising continuation method [SGBN⁺08].

The odd-number limitation has been considered in the optical control experiments in this thesis. All proof-of-concept experiments are realized with unstable states showing maximal torsion. However, during the work on this thesis the odd-number limitation was refused [FFG⁺07]. Thus, additional measurements with a torsionfree UPO are realized. Chapter 7 is concerned more deeply with the odd-number limitation and its circumvention, and presents an optical control experiment with a torsionfree orbit.

B The Amplitude Reflected by a Fabry-Perot Interferometer

Though the Fabry-Perot interferometer is textbook knowlegde, almost all studied books are concerned with the transmitted signal only. But, all-optical DFC employs the reflected signal. Thus, an expression for the signal reflected by a plane Fabry-Perot cavity is derived in this appendix. Further, in literature losses are rarely considered. In the following, two sources for losses - absorption in the cavity and absorption in the mirror coatings - are included throughout.

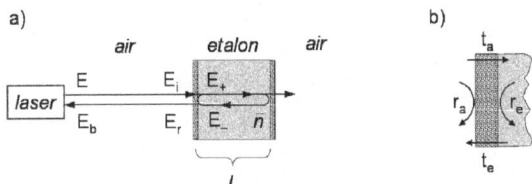

Figure B.1: *a) Sketch of FP setup with the relevant optical amplitudes. Explanations see text. b) Enlargement of a) showing the front mirror of the FP. Reflection and transmission coefficients of the mirror coating depending on the direction of the beam.*

B.1 The Control Signal in Optical DFC

At first, monochromatic plane waves with given frequency ω are assumed. E and E_b are the amplitudes emitted from and reinjected into the laser, respectively (Fig. B.1 a)). These are power amplitudes which contain a factor $\sqrt{n}$. The incident and reflected amplitude in front of the FP cavity are E_i and E_r, respectively. The FP cavity is described as an etalon with a length L, coated facets and an internal index of refraction $\bar{n}$. At the interface inside the FP cavity, a forward travelling amplitude E_+ and a backward travelling amplitude E_- are considered. In general, the mirror coatings feature transmission coefficients t_i and reflection coefficients r_i which depend on the medium where the beam is coming from, denoted by the subscripts 'e' for etalon and 'a' for air (Fig. B.1 b)). Then, the reflected amplitude E_r is given by

$$E_r = r_a E_i + t_e E_-.$$ (B.1)

The forward and backward travelling amplitudes E_+ and E_- in the FP cavity are connected via an incidence condition

$$E_+ = t_a E_i - r_e E_- \tag{B.2}$$

and a roundtrip condition

$$E_- = r_e e^{2ikL} E_+ . \tag{B.3}$$

The exponent contains the complex wavenumber of the FP etalon $k = \bar{n}\frac{\omega}{c} + i\frac{\alpha}{2}$, with α the absorption coefficient of the etalon material. Conservation of energy at the first mirror is assumed: $|E_r|^2 + |E_+|^2 = |E_i|^2 + |E_-|^2$. When including possible absorption losses A_a and A_e (subscripts refer to the medium which is left) in the mirror coatings, above equations reads

$$|E_r|^2 + |E_+|^2 = (1 - A_a)|E_i|^2 + (1 - A_e)|E_-|^2. \tag{B.4}$$

Combination of Eqs. (B.1),(B.2),(B.4) yields:

$$1 - A_a = |r_a|^2 + |t_a|^2 \tag{B.5}$$

$$1 - A_e = |r_e|^2 + |t_e|^2 \tag{B.6}$$

$$0 = t_e^* r_a + r_e^* t_a . \tag{B.7}$$

Inserting Eq. (B.2) into Eq. (B.3) yields for the backward travelling amplitude

$$E_- = \frac{r_e t_a e^{2ikL}}{1 - r_e^2 e^{2ikL}} E_i ,$$

and, thus, for the reflected amplitude

$$E_r = r_a E_i + \frac{r_e t_a t_e e^{2ikL}}{1 - r_e^2 e^{2ikL}} E_i$$

$$E_r = r_a \frac{1 - (r_e^2 - \frac{r_e t_a t_e}{r_a}) e^{2ikL}}{1 - r_e^2 e^{2ikL}} E_i$$

$$E_r = r_a \frac{1 - (1 - A_e)\frac{r_e^2}{|r_e|^2} e^{2ikL}}{1 - r_e^2 e^{2ikL}} E_i .$$

Eqs. (B.7) and (B.6) are used to transform the numerator. With $|r_e|^2 = R$ above equation becomes

$$E_r = r_a \frac{1 - (1 - A_e)\frac{r_e^2}{|r_e|^2} e^{2ikL}}{1 - R\frac{r_e^2}{|r_e|^2} e^{2ikL}} E_i ,$$

$$E_r = r_a \sum_{n=0}^{\infty} \left(R\frac{r_e^2}{|r_e|^2} e^{2ikL} \right)^n \left[1 - (1 - A_e)\frac{r_e^2}{|r_e|^2} e^{2ikL} \right] E_i .$$

So far, plane monochromatic waves with fixed frequency ω were considered. Above equation for the reflected amplitude holds for every frequency ω. A linearization of the wavenumber $k(\omega)$ is possible close to the central frequency ω_0. This yields

$$k(\omega) = k(\omega_0) + \frac{\omega - \omega_0}{v_g}$$

$$k(\omega) = k_0 + i\frac{\alpha}{2} + \frac{\omega - \omega_0}{v_g}.$$

Here, $v_g = \frac{2L}{\tau}$ is the group velocity in the FP etalon. Thus

$$\frac{r_e^2}{|r_e|^2}e^{2ikL} = e^{i\Phi_0}e^{i(\omega-\omega_0)\tau}e^{-\alpha L} = e^{i\Phi_0}e^{i(\omega-\omega_0)\tau}\mathscr{F}$$

and the expression for the reflected amplitude transforms to

$$E_r(\omega) = r_a \sum_{n=0}^{\infty} \left[Re^{i\Phi_0}\mathscr{F}e^{i(\omega-\omega_0)\tau}\right]^n \left[1 - (1-A_e)e^{i\Phi_0}\mathscr{F}e^{i(\omega-\omega_0)\tau}\right] E_i(\omega). \qquad (B.8)$$

Now, the Fourier transformation into time domain can be performed explicitly, yielding the time-dependent reflected amplitude

$$E_r(t) = \int_{-\infty}^{\infty} \frac{d\omega}{2\pi} E_r(\omega)e^{-i\omega t}$$

$$= r_a \sum_{n=0}^{\infty} \left(Re^{i(\Phi_0+\omega_0\tau)}\mathscr{F}\right)^n \left[e^{-i\omega(t-n\tau)} - (1-A_e)e^{i\Phi_0}\mathscr{F}e^{i\omega_0\tau}e^{-i\omega(t-(n+1)\tau)}\right] E_i(\omega)$$

$$= r_a \sum_{n=0}^{\infty} \left(\mathscr{R}e^{i\Phi}\right)^n \left[E_i(t-n\tau) - (1-A_e)\mathscr{F}e^{i\Phi}E_i(t-(n+1)\tau)\right]$$

with the abbreviations $\mathscr{F} = e^{-\alpha L}$, $\mathscr{R} = R\mathscr{F}$ and $\Phi = \Phi_0 + \omega_0\tau$. $E_i(t)$ and $E_r(t)$ are the slowly varying complex amplitudes of the incident and the reflected wave, respectively, in front of the etalon. $E_i(t)$ and $E_r(t)$ translate into the amplitudes $E(t)$ and $E_b(t)$ emitted by and reinjected into the laser, respectively, by two linear relations:

$$E_b(t) \sim E_r(t - \frac{\tau_l}{2})$$

and

$$E_i(t) \sim E(t - \frac{\tau_l}{2})$$

with τ_l the group roundtrip time between laser and FP etalon. The proportionality factors are complex and contain phase shifts and coupling efficiencies of the respective setup. They can be combined with r_a in one prefactor $Ke^{i\varphi}$.

Thus, one finally obtains for the reflected amplitude $E_b(t)$ in optical DFC

$$E_b(t) = -Ke^{i\varphi} \sum_{n=0}^{\infty} (\mathscr{R}e^{i\Phi})^n \left[(1-A_e)\mathscr{F}e^{i\Phi} E(t_{n+1}) - E(t_n) \right] \tag{B.9}$$

with

$$t_n = t - \tau_l - n\tau.$$

B.2 The Intensity Transmitted by the FP

In the experiment, the accessible variable is the intensity $I_t = \frac{E_t \cdot E_t^*}{2}$ transmitted by the FP. By measuring the dependence of the transmitted intensity on the wavelength of the incident cw emission, a transmission spectrum $I_t(\lambda)$ can be derived, which enables a determination of the actual FP parameters. The amplitude E_t transmitted by the FP is

$$E_t = t_e E_+ . \tag{B.10}$$

The forward travelling amplitude E_+ follows from (B.3) and (B.2) to

$$E_+ = \frac{t_a}{1 - r_e^2 e^{2ikL}} E_i .$$

This yields for the transmitted amplitude

$$E_t = t_e E_+ = \frac{t_a t_e}{1 - r_e^2 e^{2ikL}} E_i ,$$

and thus

$$\frac{E_t}{E_i} = \frac{\sqrt{1 - r_a^2 - A_a}\sqrt{1 - r_e^2 - A_e}}{1 - r_e^2 e^{2ikL}} .$$

Assuming $r_e = r_a$ and $A_e = A_a$ above expression simplifies to

$$\frac{E_t}{E_i} = \frac{1 - r^2 - A}{1 - r^2 e^{2ikL}} .$$

With $\delta = \frac{2\bar{n}\omega L}{c}$ this yields for the transmitted amplitude

$$\frac{E_t}{E_i} = \frac{1 - r^2 - A}{1 - r^2 e^{i\delta} e^{-\alpha L}} \tag{B.11}$$

and for the transmitted intensity

$$\frac{I_t}{I_i} = \frac{(1 - R - A)^2}{1 + R^2 \cdot e^{-2\alpha L} - 2R \cdot \cos\delta \cdot e^{-\alpha L}} . \tag{B.12}$$

The FP etalons used in the experiments come with different lenghts L and mirror reflectivies R. The values for L and R are given by the producer within a certain range. In order to determine the FP parameters L, R, A, α and $\bar{n}$ with sufficient certainty, transmission spectra $I_t/I_i(\lambda)$ are measured (Sec. 2.2), and Eq. (B.12) is fitted to the data.

C Noninvasive Control and Noise in Literature

In DFC, the control signal vanishes ideally as the systems reaches the goal dynamics, thereby enabling noninvasive control. However, in a real system the dynamics will never converge exactly to the target state, because noise continuously drives it away. The target state is only controlled "incompletely". This effect comes along with a not completely vanishing control signal, which goes to a minimal, but finite value. The question arises, especially with experimental realizations, whether one can call such control processes still "noninvasive".

The aspects of imperfect control, and of a nonvanishing control signal are in general not discussed deeply in the reviewed experimental literature [PT93, BR99, BDG94, DTH98, CMA95]. Often, only a value for the level of the remaining control signal is given. In the first experimental realization of DFC by Pyragas and Tamasevicius [PT93] with an electrical circuit, the minimal control signal level is well below 1 %, and not discussed further. Independently from Pyragas work, the DFC method has been presented and experimentally verified by Bielawski and coworkers in [BDG94] using a CO_2 laser with modulated losses. Here, the control signal under stabilization is below a level of 2%, consisting of a periodic part and an irregular part. The authors discuss the question, whether the stabilized orbit can be considered identical to the unstable orbit of the unperturbed system. In [DTH98], the first realization of delayed feedback control of an autonomous system is demonstrated, using a single-mode optically pumped laser showing Lorenz-like chaos. The control signal level here ranges from 0.1 to 5 %. In the discussion, the authors distinguish between "perturbative" and "not perturbative" control. In [BR99], control of a spin system by different control methods including time-delayed feedback is presented. The suppression of chaos by time-delayed feedback is apparently incomplete, while the control signal is below a noise level of 1%. The question of incomplete control is not discussed here.

An interesting paper in this context is [CMA95]. Here, the authors develop a control method which they see intermediate between OGY, DFC and a non-feedback method - the control of chaos by small modulations of a control parameter. The latter is, of course, an invasive method. In [CMA95], the necessary strength of the modulations is minimized by choosing as modulation frequencies only those corresponding to the highest peaks of the power spectrum - stabilizing only the leading cycles of the chaotic attractor. This way, they end up with a minimal level of perturbation of 2% - which is of the same order as the level of the control signal in, e.g., [BDG94] and [DTH98].

In the present work, an upper border for the residual control signal is estimated for each control experiment, the values ranging between 10^{-5} and 10^{-3} of the laser output. The question, if the control can be considered 'noninvasive' here, is tested with additional experiments, where the FP is replaced by a simple mirror providing the same feedback level (Chapter 6). First, it is found that simple mirror feedback of the order of the residual control signal does not lead to a change of the mode of operation in the laser. Thus, the control effect in the experiments is not due to a shift of the point of operation by the residual feedback. Second, simple mirror feedback of the order of the residual control power does affect the target state, but only extremely weakly (a frequency shift of less than 1 per mille is found for a mirror reflectivity of 10^{-3}).

D Publications and Presentations

Peer-reviewed Publikations and Book Chapters:

1. S. Schikora, P. Hövel, H.-J. Wünsche, E. Schöll and F. Henneberger. All-optical noninvasive control of unstable steady states in a semiconductor laser. Phys. Rev. Lett. 97, 213902 (2006).

2. S. Schikora, H.-J. Wünsche and F. Henneberger. All-optical noninvasive control of semiconductor lasers. Proc. SPIE 6889. Physics and Simulation of Optoelectronic Devices XVI, 688909 (2008).

3. H.-J. Wünsche, S. Schikora and F. Henneberger. Noninvasive Control of Semiconductor Lasers by Delayed Optical Feedback. in: Handbook of Chaos Control. Wiley-VCH, Weinheim, 2nd edition, Chapter 21 (2008).

4. S. Schikora, H.-J. Wünsche and F. Henneberger. All-optical noninvasive chaos control of a semiconductor laser. Phys. Rev. E 78, 025202(R) (2008).

5. T. Dahms, V. Flunkert, F. Henneberger, P. Hövel, S. Schikora, E. Schöll, and H.-J. Wünsche. Noninvasive optical control of complex semiconductor laser dynamics. Eur. Phys. J. Special Topics 191, 71-89 (2010).

6. S. Schikora, H.-J. Wünsche and F. Henneberger. Odd-number theorem: Optical feedback control at a subcritical Hopf bifurcation in a semiconductor laser. Phys. Rev. E 83(2), 026203 (2011).

Presentations at international Conferences and Workshops:

1. S. Schikora, All-optical control of a semiconductor laser by time-delayed feedback: experiment and simulation, Workshop "Complex Dynamics and Delay Effects in Coupled Systems", Weierstrass Institute for Applied Analysis and Stochastic (WIAS), September 11th - 13th 2006, Berlin.

2. S. Schikora, H.-J. Wünsche, and F. Henneberger, Delayed optical feedback control of semiconductor lasers: experiment and simulation, Symposium "Delay und Kontrolle" SFB 555, April 27th 2007, Berlin.

3. S. Schikora, H.-J. Wünsche, and F. Henneberger, All-optical time-delayed feedback control of semiconductor lasers, CLEO/Europe-IQEC 2007, June 17th - 22nd 2007, München.

4. S. Schikora, P. Hövel, H.-J. Wünsche, E. Schöll, and F. Henneberger, All-optical noninvasive control of unstable steady states in a semiconductor laser, Poster, 3rd International IEEE Scientific Conference on Physics and Control (PhysCon 2007), September 3rd - 7th 2007, Potsdam.

5. S. Schikora, H.-J. Wünsche, and F. Henneberger, All-optical chaos control of multisection lasers, Workshop "Nonlinear Dynamics in Semiconductor Lasers", Weierstrass Institute for Applied Analysis and Stochastic (WIAS), November 19th - 21st 2007, Berlin, invited.

6. S. Schikora, H.-J. Wünsche, and F. Henneberger, All-optical noninvasive control of semiconductor lasers, Photonics Europe 2008, April 7th -10th 2008, Strasbourg, France.

7. S. Schikora, H.-J. Wünsche, and F. Henneberger, Delayed optical feedback control at a subcritical Hopf bifurcation in a semiconductor laser, Dynamics Days Berlin - Brandenburg 2008, October 8th - 10th, 2008, Potsdam.

Bibliography

[AB65]   ARECCHI, F. T. and R. BONIFACI: *Theory of optical maser amplifiers.* IEEE J. Quant. Electron., 1:169, 1965.

[ACacE+87] AUERBACH, DITZA, PREDRAG CVITANOVIĆ, JEAN-PIERRE ECKMANN, GEMUNU GUNARATNE and ITAMAR PROCACCIA: *Exploring chaotic motion through periodic orbits.* Phys. Rev. Lett., 58(23):2387–2389, 1987.

[AD93]   AGRAWAL, G. P. and N. K. DUTTA: *Semiconductor Lasers* , chapter 6. Springer, 2nd edition, 1993.

[AMKK+07]   AL-MUMIN, MOHAMMED, CHEOLHWAN KIM, INWOONG KIM, NAZAR JAAFAR and GUIFANG LI: *Injection locked multi-section gain-coupled dual mode DFB laser for terahertz generation.* Opt. Comm., 275(1):186–189, 2007.

[AP04]   AHLBORN, ALEXANDER and ULRICH PARLITZ: *Stabilizing Unstable Steady States Using Multiple Delay Feedback Control.* Phys. Rev. Lett., 93(26):264101, 2004.

[ASL+05]   ARGYRIS, A., D. SYVRIDIS, L. LARGER, V. ANNOVAZZI-LODI, P. COLET, I. FISCHER, J. GARCIA-OJALVO, C. R. MIRASSO, L. PESQUERA and K. A. SHORE: *Chaos-based communications at high bit rates using commercial fiber-optic links.* Nature, 438:343, 2005.

[BAM04]   BOCCALETTI, S., E. ALLARIA and R. MEUCCHI: *Experimental control of coherence of a chaotic oscillator.* Phys. Rev. E, 69:066211, 2004.

[BAS+02]   BECK, O., A. AMANN, E. SCHÖLL, J. E. S. SOCOLAR and W. JUST: *Comparison of time-delayed feedback schemes for spatiotemporal control of chaos in a reaction-diffusion system with global coupling.* Phys. Rev. E, 66(1):016213, 2002.

[Bau04]   BAUER, STEFAN: *Nonlinear dynamics of semiconductor lasers with active optical feedback.* PhD thesis, Humboldt-University Berlin, Department of Physics, 2004.

[BBDG93]   BIELAWSKI, S., M. BOUAZAOUI, D. DEROZIER and P. GLORIEUX: *Stabilization and characterization of unstable steady states in a laser.* Phys. Rev. A, 47(4):3276–3279, 1993.

[BBK+02]   BAUER, S., 0. BROX, J. KREISSL, G. SAHIN and B. SARTORIUS: *Optical microwave source.* Electron. Lett., 38:334, 2002.

[BBK⁺04]   BAUER, S., O. BROX, J. KREISSL, B. SARTORIUS, M. RADZIUNAS,
            J. SIEBER, H.-J. WÜNSCHE and F. HENNEBERGER: *Nonlinear Dynam-
            ics of Semiconductor Lasers with Active Optical Feedback.* Phys. Rev. E,
            69:016206, 2004.

[BBM⁺03]   BETA, C., M. BERTRAM, A. S. MIKHAILOV, H. H. ROTERMUND and
            G. ERTL: *Controlling turbulence in a surface chemical reaction by time-
            delay autosynchronization.* Phys. Rev. E, 67:046224, 2003.

[BDG94]   BIELAWSKI, S., D. DEROZIER and P. GLORIEUX: *Controlling unstable
            periodic orbits by a delayed continuous feedback.* Phys. Rev. E, 49:R971–
            R974, 1994.

[BGL⁺00]   BOCCALETTI, S., C. GREBOGI, Y.-C. LAI, H. MANCINI and D. MAZA:
            The control of chaos: theory and applications. Physics Reports, 329:103–
            197, 2000.

[BHL89]   BAUGHER, A., P. HAMMACK and J. LIN: *Subcritical Hopf bifurcation in a
            NMR laser with an injected signal.* Phys. Rev. A, 39(3):1549–1551, 1989.

[BHS99]   BLASIUS, B., A. HUPPERT and L. STONE: *Complex dynamics and phase
            synchronization in spatially extended ecological systems.* Nature, 399:354,
            1999.

[BIG04]   BLAKELY, J. N., L. ILLING and D. GAUTHIER: *Controlling fast chaos in
            delay dynamical systems.* Phys. Rev. Lett., 92(193901), 2004.

[BJS04]   BALANOV, A. G., N. B. JANSON and E. SCHÖLL: *Control of noise-
            induced oscillations by delayed feedback.* Physica D, 199:1–12, 2004.

[BPS10]   BROWN, G., C. POSTLETHWAITE and M. SILBER: *Time-delayed feed-
            back control of unstable periodic orbits near a subcritical Hopf bifurcation.*
            arXiv:1006.3479v1, 2010.

[BR99]   BENNER, H. and E. REIBOLD: *Handbook of Chaos Control.* Wiley-VCH,
            Weinheim, 1st edition, 1999.

[Bro05]   BROX, OLAF: *DFB-Laser mit integriert optischer Rueckkopplung fuer die
            optische Signalverarbeitung.* PhD thesis, Technical University Berlin, 2005.

[BS96]   BLEICH, M. E. and J. E. S. SOCOLAR: *Stability of periodic orbits con-
            trolled by time-delay feedback.* Phys. Lett. A, 210(1-2):87, 1996.

[BSS04]   BORNHOLDT, C., J. SLOVAK and B. SARTORIUS: *Semiconductor-based
            all-optical 3R regenerator demonstrated at 40 Gbit/s.* Electron. Lett.,
            40:192, 2004.

[CBH⁺98]   CHANG, A., J. C. BIENFANG, G. M. HALL, J. R. GARDNER and D. J.
            GAUTHIER: *Stabilizing unstable steady states using extended time-delay
            autosynchronization.* Chaos, 8(4):782, 1998.

[CLMG99] CIOFINI, M., A. LABATE, R. MEUCCI and M. GALANTI: *Stabilization of unstable fixed points in the dynamics of a laser with feedback*. Phys. Rev. E, 60(1):398–402, 1999.

[CMA95] CIOFINI, M., R. MEUCCI and F. T. ARECCHI: *Experimental control of chaos in a laser*. Phys. Rev. E, 52:94, 1995.

[CRW94] COLET, PERE, RAJARSHI ROY and KURT WIESENFELD: *Controlling hyperchaos in a multimode laser model*. Phys. Rev. E, 50(5):3453–3457, 1994.

[Cry02] *Feature section on optical chaos and applications to cryptography*. IEEE Journ. Quant. Electron., 38(9):1138–1204, 2002.

[CS95] CARR, THOMAS W. and IRA B. SCHWARTZ: *Controlling the unstable steady state in a multimode laser*. Phys. Rev. E, 51(5):5109–5111, 1995.

[DFH⁺10] DAHMS, T., V. FLUNKERT, F. HENNEBERGER, P. HÖVEL, S. SCHIKORA, E. SCHÖLL and H-J. WÜNSCHE: *Noninvasive optical control of complex semiconductor laser dynamics*. Eur. Phys. J. Special Topics, 191(2):71–89, 2010.

[DHD87] DAHMANI, B., L. HOLLBERG and R. DRULLINGER: *Frequency stabilization of semiconductor lasers by resonant optical feedback*. Opt. Lett., 12(11):876–878, 1987.

[DHK⁺83] DREVER, R. W. P., J. L. HALL, F. V. KOWALSKI, J. HOUGH, G. M. FORD, A. J. MUNLEY and H. WARD: *Laser phase and frequency stabilization using an optical resonator*. Appl. Phys. B, 31(2):97–105, 1983.

[DHS07] DAHMS, THOMAS, PHILIPP HOVEL and ECKEHARD SCHOLL: *Control of unstable steady states by extended time-delayed feedback*. Phys. Rev. E, 76(5):056201, 2007.

[DTH98] DYKSTRA, R., D. Y. TANG and N. R. HECKENBERG: *Experimental control of single-mode laser chaos by using continuous, time-delayed feedback*. Phys. Rev. E, 57(6):6596–6598, 1998.

[EJ09] ERZGRÄBER, H. and W. JUST: *Global view on a nonlinear oscillator subject to time-delayed feedback control*. Physica D, 238(16):1680–1687, 2009.

[EKL⁺06] ERZGRABER, H., B. KRAUSKOPF, D. LENSTRA, A. P. A. FISCHER and G. VEMURI: *Frequency versus relaxation oscillations in a semiconductor laser with coherent filtered optical feedback*. Phys. Rev. E., 73(5):055201, 2006.

[ELK⁺07] ERZGRABER, H., D. LENSTRA, B. KRAUSKOPF, A. P. A. FISCHER and G. VEMURI: *Feedback phase sensitivity of a semiconductor laser subject to filtered optical feedback: Experiment and theory*. Phys. Rev. E, 76(2):026212, 2007.

[ER85] ECKMANN, J.-P. and D. RUELLE: *Ergodic theory of chaos and strange attractors*. Rev. Mod. Phys., 57:617, 1985.

[FAY+00] FISCHER, A.P.A., O.K. ANDERSEN, M. YOUSEFI, S. STOLTE and D. LENSTRA: *Experimental and theoretical study of filtered optical feedback in a semiconductor laser.* IEEE Journ. Quant. Electron., 36(3):375–384, 2000.

[FBS99] FRANCESCHINI, G., S. BOSE and E. SCHÖLL: *Control of chaotic spatiotemporal spiking by time-delay autosynchronization.* Phys. Rev. E, 60(5):5426–5434, 1999.

[FFG+07] FIEDLER, B., V. FLUNKERT, M. GEORGI, P. HOVEL and E. SCHOLL: *Refuting the Odd-Number Limitation of Time-Delayed Feedback Control.* Phys. Rev. Lett., 98(11):114101, 2007.

[Fle68] FLECK, J. A.: *Emission of Pulse Trains by Q-Switched Lasers.* Phys. Rev. Lett., 21:131, 1968.

[FMD85] FARQUHAR, R., D. MUHONEN and S. A. DAVIS: *Trajectories and Orbit Maneuvers for the ISEE-3/ICE Comet Mission.* J. Astronaut. Sci., 33:235, 1985.

[FS07] FLUNKERT, V. and E. SCHÖLL: *Suppressing noise-induced intensity pulsations in semiconductor lasers by means of time-delayed feedback.* Phys. Rev. E, 76(6):066202, 2007.

[FYF+08] FIEDLER, B., S. YANCHUK, V. FLUNKERT, P. HÖVEL, H.-J. WÜNSCHE and E. SCHÖLL: *Delay stabilization of rotating waves near fold bifurcation and application to all-optical control of a semiconductor laser.* Phys. Rev. E, 77(6):066207, 2008.

[FYL+04] FISCHER, ALEXIS P. A., MIRVAIS YOUSEFI, DAAN LENSTRA, MICHAEL W. CARTER and GAUTAM VEMURI: *Filtered Optical Feedback Induced Frequency Dynamics in Semiconductor Lasers.* Phys. Rev. Lett., 92(2):023901, Jan 2004.

[Gau98] GAUTHIER, D. J.: *Controlling lasers by use of extended time-delay autosynchronization.* Opt. Lett., 23(9):703, 1998.

[GCB02] GROSSE, P., M. J. CASSIDY and P. BROWN: *EEG-EMG, MEG-EMG and EMG-EMG frequency analysis: physiological principles and clinical applications.* Clin. Neurophysiol., 113(10):1523–1531, 2002.

[Gio91] GIONA, M.: *Dynamics and relaxation properties of complex systems with memory.* Nonlinearity, 4:911–925, 1991.

[GIR+92] GILLS, ZELDA, CHRISTINA IWATA, RAJARSHI ROY, IRA B. SCHWARTZ and IOANA TRIANDAF: *Tracking unstable steady states: Extending the stability regime of a multimode laser system.* Phys. Rev. Lett., 69(22):3169–3172, 1992.

[GRP03] GOLDOBIN, D., M. ROSENBLUM and A. PIKOVSKY: *Controlling oscillator coherence by delayed feedback.* Phys. Rev. E, 67(6):061119, 2003.

[GSCS94]   GAUTHIER, D. J., D. W. SUKOW, H. M. CONCANNON and J. E. S. SO-
            COLAR: *Stabilizing unstable periodic orbits in a fast diode resonator us-
            ing continuous time-delay autosynchronization.* Phys. Rev. E, 50(3):2343–
            2346, 1994.

[GW00]   GUCKENHEIMER, J. and A. R. WILLMS: *Asymptotic analysis of subcriti-
            cal Hopf-homoclinic bifurcation.* Physica D, 139:195–216, 2000.

[HHD+10]   HOU, LIANPING, MOHSIN HAJI, RAFAL DYLEWICZ, PIOTR STOLARZ,
            BOCANG QIU, EUGENE A. AVRUTIN and A. CATRINA BRYCE: *160
            GHz harmonic mode-locked AlGaInAs 1.55 mu m strained quantum-well
            compound-cavity laser.* Opt. Lett., 35(23):3991–3993, 2010.

[HMM96]   HOMAR, M., J. V. MOLONEY and M. SAN MIGUEL: *Traveling wave
            model of a multimode Fabry-Perot laser in free running and external cavity
            configurations.* IEEE J. Quant. Electron., 32:553–566, 1996.

[HS03]   HOEVEL, P. and J. E. S. SOCOLAR: *Stability domains for time-delayed
            feedback control with latency.* Phys. Rev. E, 68:036206, 2003.

[HS05]   HOEVEL, P. and E. SCHOELL: *Control of unstable steady states by time-
            delayed feedback methods.* Phys. Rev. E, 72:046203, 2005.

[HSC+07]   HÖHNE, KLAUS, HIROYUKI SHIRAHAMA, CHOL-UNG CHOE, HARTMUT
            BENNER, KESTUTIS PYRAGAS and WOLFRAM JUST: *Global Properties
            in an Experimental Realization of Time-Delayed Feedback Control with an
            Unstable Control Loop.* Phys. Rev. Lett., 98(21):214102, 2007.

[HU00]   HOLYST, J. A. and K. URBANOWICZ: *Chaos control in economical model
            by time-delayed feedback method.* Physica A, 287:587–598, 2000.

[Hun91]   HUNT, E. R.: *Stabilizing high-period orbits in a chaotic system: The diode
            resonator.* Phys. Rev. Lett., 67(15):1953–1955, 1991.

[JBO+97a]   JUST, W., T. BERNARD, M. OSTHEIMER, E. REIBOLD and H. BEN-
            NER: *Mechanism of Time-Delayed Feedback Control.* Phys. Rev. Lett.,
            78(2):203–206, Jan 1997.

[JBO+97b]   JUST, WOLFRAM, THOMAS BERNARD, MATTHIAS OSTHEIMER, EKKE-
            HARD REIBOLD and HARTMUT BENNER: *Mechanism of Time-Delayed
            Feedback Control.* Phys. Rev. Lett., 78(2):203–206, 1997.

[JBS04]   JANSON, N. B., A. G. BALANOV and E. SCHOELL: *Delayed Feed-
            back as a Means of Control of Noise-Induced Motion.* Phys. Rev. Lett.,
            93(1):010601, 2004.

[JFG+07]   JUST, W., B. FIEDLER, M. GEORGI, V. FLUNKERT, P. HOVEL and
            E. SCHÖLL: *Beyond the odd number limitation: A bifurcation analysis of
            time-delayed feedback control.* Phys. Rev. E, 76(2):026210, 2007.

[JRRB99]   JUST, W., D. RECKWERTH, E. REIBOLD and H. BENNER: *Influence of control loop latency on time-delayed feedback control.* Phys. Rev. E, 59:2826, 1999.

[JTH93]   JOHNSON, G. A., T. E. TIGNER and E. R. HUNT: *Controlling chaos in Chua's circuit.* J. Circuits Syst. Comput., 3:119, 1993.

[KHF⁺09]   KEHRT, M., P. HÖVEL, V. FLUNKERT, M. A. DAHLEM, P. RODIN and E. SCHÖLL: *Stabilization of complex spatio-temporal dynamics near a subcritical Hopf bifurcation by time-delayed feedback.* Eur. Phys. J. B, 68:557–565, 2009.

[KMH09]   KORONOVSKII, A. A., O. I. MOSKALENKO and A. E. HRAMOV: *On the use of chaotic synchronization for secure communication.* Physics-Uspekhi, 52(12):1213–1238, 2009.

[LCB89]   LAURENT, PH., A. CLAIRON and CH. BREANT: *Frequency noise analysis of optically self-locked diode lasers.* IEEE Journ. Quant. Electron., 25(6):1131, 1989.

[LGWH09]   LOOSE, A., B. K. GOSWAMI, H.-J. WÜNSCHE and F. HENNEBERGER: *Tristability of a semiconductor laser due to time-delayed optical feedback.* Phys. Rev. E, 79(3):036211, 2009.

[LH94]   LU, W. and R. G. HARRISON: *Controlling chaos using continuous interference feedback: proposal for all-optical devices.* Opt. Comm., 109:457–461, 1994.

[LK80]   LANG, R. and K. KOBAYASHI: *External optical feedback effects on semiconductor injection laser properties.* IEEE Journ. Quant. Electron., 16(3):347, 1980.

[LRL⁺07]   LAVIGNE, B., J. RENAUDIER, F. LELARGE, O. LEGOUEZIGOU, H. GARIAH and G.-H. DUAN: *Polarization-insensitive low timing jitter and highly optical noise tolerant all-optical 40-GHz clock recovery using a bulk and a quantum-dots-based self-pulsating laser cascade.* Journ. Lightwave Technol., 25(1):170, 2007.

[LVdB85]   LENSTRA, D., B. VERBEEK and A. J. DEN BOEF: *Coherence collapse in single-mode semiconductor lasers due to optical feedback.* IEEE Journ. Quant. Electron., 21(6):674, 1985.

[LWP01]   LUETHJE, O., S. WOLFF and G. PFISTER: *Control of chaotic Taylor-Couette flow with time-delayed feedback.* Phys. Rev. Lett, 86(9):1745–1748, 2001.

[LY75]   LI, T. Y. and J. YORKE: *Period three implies chaos.* The American Mathematical Monthly, 82(10):985–992, 1975.

[Max68]   MAXWELL, J. C.: *On Governors.* Proceedings of the Royal Society of London, 16:270–283, 1868.

[MCA96]   MEUCCI, R., M. CIOFINI and R. ABBATE: *Suppressing chaos in lasers by negative feedback*. Phys. Rev. E, 53(6):R5537–R5540, 1996.

[MGA94]   MEUCCI, R., W. GADOMSKI and F. T. ARECCHI: *Experimental control of chaos by means of weak parametric perturbations*. Phys. Rev. E, 49:R2528–R2531, 1994.

[Nak97]   NAKAJIMA, H.: *On analytical properties of delayed feedback control of chaos*. Phys. Lett. A, 232:207–210, 1997.

[NKC04]   NUMATA, KENJI, AMY KEMERY and JORDAN CAMP: *Thermal-Noise Limit in the Frequency Stabilization of Lasers with Rigid Cavities*. Phys. Rev. Lett., 93:250602, 2004.

[NPT95]   NAMAJUNAS, A., K. PYRAGAS and A. TAMASEVICIUS: *Stabilization of an unstable steady state in a Mackey-Glass system*. Phys. Lett. A, 204:255, 1995.

[NSS97]   NEIMAN, A., P. I. SAPARIN and L. STONE: *Coherence resonance at noisy precursors of bifurcations in nonlinear dynamical systems*. Phys. Rev. E, 56(1):270–273, 1997.

[NU98a]   NAKAJIMA, HIROYUKI and YOSHISUKE UEDA: *Half-period delayed feedback control for dynamical systems with symmetries*. Phys. Rev. E, 58(2):1757–1763, 1998.

[NU98b]   NAKAJIMA, HIROYUKI and YOSHISUKE UEDA: *Limitation of generalized delayed feedback control*. Physica D: Nonlinear Phenomena, 111(1-4):143 – 150, 1998.

[OGY90]   OTT, E., C. GREBOGI and J. YORKE: *Controlling chaos*. Phys. Rev. Lett., 64:1196, 1990.

[Ott]   OTT, E.: *Chaos in dynamical systems*.

[PAS05]   POMPLUN, J., A. AMANN and E. SCHÖLL: *Mean-field approximation of time-delayed feedback control of noise-induced oscillations in the Van der Pol system*. Europhys. Lett., 71(3):366, 2005.

[PBA96]   PIERRE, T., G. BONHOMME and A. ATIPO: *Controlling the chaotic regime of nonlinear ionization waves using the time-delay autosynchronization method*. Phys. Rev. Lett, 76:2290, 1996.

[PFA+92]   PAPOFF, F., A. FIORETTI, E. ARIMONDO, G. B. MINDLIN, H. SOLARI and R. GILMORE: *Structure of chaos in the laser with saturable absorber*. Phys. Rev. Lett., 68(8):1128–1131, Feb 1992.

[PMR+99a] PARMANANDA, P., R. MADRIGAL, M. RIVERA, L. NYIKOS, I. Z. KISS and V. GÁSPÁR: *Stabilization of unstable steady states and periodic orbits in an electrochemical system using delayed-feedback control*. Phys. Rev. E, 59(5):5266–5271, 1999.

[PMR+99b]   PARMANANDA, P., R. MADRIGAL, M. RIVERA, L. NYIKOS, I. Z. KISS
and V. GASPAR: *Stabilization of unstable steady states and periodic orbits
in an electrochemical system using delayed-feedback control.* Phys. Rev. E,
59(5):5266–5271, 1999.

[POS+83]   PATZAK, E., H. OLESEN, A. SAGIMURA, S. SAITO and T. MUKAI: *Spectral linewidth reduction in semiconductor lasers by an external cavity with
weak optical feedback.* Electron. Lett., 19:938–940, 1983.

[PP06a]   PAWLIK, A. H. and A. PIKOVSKY: *Control of oscillators coherence by
multiple delayed feedback.* Phys. Lett. A, 358:181–185, 2006.

[PP06b]   PYRAGAS, V. and K. PYRAGAS: *Delayed feedback control of the Lorenz
system: An analytical treatment at a subcritical Hopf bifurcation.* Phys.
Rev. E, 73(3):036215, 2006.

[PPB04]   PYRAGAS, K., V. PYRAGAS and H. BENNER: *Delayed feedback control of dynamical systems at a subcritical Hopf bifurcation.* Phys. Rev. E,
70(5):056222, 2004.

[PPKH02]   PYRAGAS, K., V. PYRAGAS, I. Z. KISS and J. L. HUDSON: *Stabilizing
and Tracking Unknown Steady States of Dynamical Systems.* Phys. Rev.
Lett., 89(24):244103, 2002.

[PRK01]   PIKOVSKY, A., M. ROSENBLUM and J. KURTHS: *Synchronization. A Universal Concept in Nonlinear Sciences.* University Press, Cambridge, 2001.

[PRW+06]   PEREZ, T., M. RADZIUNAS, H.-J. WÜNSCHE, C. R. MIRASSO and
F. HENNEBERGER: *Synchronization properties of two coupled multisection
semiconductor lasers emitting chaotic light.* IEEE Photon. Techn. Lett.,
18(20):2135–2137, 2006.

[PS07]   POSTLETHWAITE, C. M. and M. SILBER: *Stabilizing unstable periodic orbits in the Lorenz equations using time-delayed feedback control.*
Phys. Rev. E, 76(5):056214, 2007.

[PT93]   PYRAGAS, K. and A. TAMASEVICIUS: *Experimental control of chaos by
delayed self-controlling feedback.* Phys. Lett. A, 180:99–102, 1993.

[Pyr92]   PYRAGAS, K.: *Continuous control of chaos by self-controlling feedback.*
Phys. Lett. A, 170:421, 1992.

[Pyr01]   PYRAGAS, K.: *Control of Chaos via an Unstable Delayed Feedback Controller.* Phys. Rev. Lett., 86(11):2265–2268, 2001.

[Pyr06]   PYRAGAS, K.: *Delayed feedback control of chaos.* Philos. Trans. R. Soc.
London A, 364:2309, 2006.

[Rad06]   RADZIUNAS, M.: *Numerical bifurcation analysis of the traveling wave
model of multisection semiconductor lasers.* Physica D, 213:98, 2006.

[Ran82]   RAND, D.: *Dynamics and Symmetry. Predictions for modulated waves in rotating fluids.* Arch. Rat. Mech. Analysis, 79:1–38, 1982.

[RDLG07] RENAUDIER, J., G.-H. DUAN, P. LANDAIS and PH. GALLION: *Phase correlation and linewidth reduction of 40 GHz self-pulsation in distributed Bragg reflector semiconductor laser.* IEEE Journ. Quant. Electron., 43(2):147, 2007.

[RFBB93] REYL, C., L. FLEPP, R. BADII and E. BRUN: *Control of NMR-laser chaos in high-dimensional embedding space.* Phys. Rev. E, 47(1):267–272, 1993.

[RMD⁺99] ROGISTER, F., P. MÉGRET, O. DEPARIS, M. BLONDEL and T. ERNEUX: *Suppression of low-frequency fluctuations and stabilization of a semiconductor laser subjected to optical feedback from a double cavity: theoretical results.* Opt. Lett., 24(17):1218–1220, 1999.

[RMM⁺92] ROY, RAJARSHI, T. W. MURPHY, T. D. MAIER, Z. GILLS and E. R. HUNT: *Dynamical control of a chaotic laser: Experimental stabilization of a globally coupled system.* Phys. Rev. Lett., 68(9):1259–1262, 1992.

[RN01]   RAMESH, M. and S. NARAYANAN: *Controlling Chaotic Motions in a two-dimensional airfoil using time-delayed feedback.* Journ. Sound Vibr., 239(5):1037–1049, 2001.

[RP04]   ROSENBLUM, M. and A. PIKOVSKY: *Delayed feedback control of collective synchrony: An approach to suppression of pathological brain rhythms.* Phys. Rev. E, 70(4):041904), 2004.

[RV77]   RISCH, C. H. and C. VOUMARD: *Self-pulsation in the output intensity and spectrum of GaAs-AlGaAs cw diode lasers coupled to a frequency-selective external optical cavity.* Journ. Appl. Phys., 48:2083, 1977.

[SBGS97] SUKOW, D.W., M. E. BLEICH, D. J. GAUTHIER and J. E. S. SOCOLAR: *Controlling chaos in a fast diode resonator using extended time-delay autosynchronization: Experimental observations and theoretical analysis.* Chaos, 7:560, 1997.

[Sch99]   SCHUSTER, H. G. (editor): *Handbook of Chaos Control.* Wiley-VCH, Weinheim, 1st edition, 1999.

[SG98]   SOCOLAR, JOSHUA E. S. and DANIEL J. GAUTHIER: *Analysis and comparison of multiple-delay schemes for controlling unstable fixed points of discrete maps.* Phys. Rev. E, 57(6):6589–6595, 1998.

[SGBN⁺08] SIEBER, J., A. GONZALEZ-BUELGA, S. A. NEILD, D. J. WAGG and B. KRAUSKOPF: *Experimental Continuation of Periodic Orbits through a Fold.* Phys. Rev. Lett., 100(24):244101, 2008.

[SH96]   SIMMENDINGER, C. and O. HESS: *Controlling delay-induced chaotic behavior of a semiconductor laser with optical feedback.* Phys. Lett. A, 216:97–105, 1996.

[SHW⁺06] SCHIKORA, S., P. HOEVEL, H.-J. WUENSCHE, E. SCHOELL and F. HEN-
NEBERGER: *All-optical noninvasive control of unstable steady states in a semiconductor laser*. Phys. Rev. Lett., 97(213902), 2006.

[Sie02] SIEBER, J.: *Numerical bifurcation analysis for multi-section semiconductor lasers*. SIAM J. of Appl. Dyn. Sys., 1(2):248–270, 2002.

[Slo05] SLOVAK, J. In *Proc. 31th Europ. Conf. on Optical Communication (ECOC)*, Glasgow (UK), 2005.

[SM06] SERRAT, C. and C. MASOLLER: *Modeling spatial effects in multi-longitudinal-mode semiconductor lasers*. Phys. Rev. A, 73:043812, 2006.

[SRKA98] SCHÄFER, C., M. G. ROSENBLUM, J. KURTHS and H.-H. ABEL: *Heart-beat synchronized with ventilation*. Nature, 392:239, 1998.

[SRS] SIEBER, J., M. RADZIUNAS and K.R. SCHNEIDER: *Dynamics of multi-section lasers*. Mathematical Modelling and Analysis, 9.

[SS97] SCHUSTER, H. G. and M. B. STEMMLER: *Control of chaos by oscillating feedback*. Phys. Rev. E, 56(6):6410–6417, 1997.

[SS08] SCHÖLL, E. and H. G. SCHUSTER (editors): *Handbook of Chaos Control*. Wiley-VCH, Weinheim, 2nd edition, 2008.

[SSG94] SOCOLAR, J. E. S., D. W. SUKOW and D. J. GAUTHIER: *Stabilizing unstable periodic orbits in fast dynamical systems*. Phys. Rev. E, 50:3245, 1994.

[SWB91] SINGER, J., Y.-Z. WANG and HAIM H. BAU: *Controlling a chaotic system*. Phys. Rev. Lett., 66(9):1123–1125, 1991.

[SWH08] SCHIKORA, S., H.-J. WUENSCHE and F. HENNEBERGER: *All-optical noninvasive chaos control of a semiconductor laser*. Phys. Rev. E, 78(025202(R)), 2008.

[SWH11] SCHIKORA, S., H-J. WÜNSCHE and F. HENNEBERGER: *Odd-number theorem: Optical feedback control at a subcritical Hopf bifurcation in a semiconductor laser*. Phys. Rev. E, 83(2):026203, 2011.

[Tas02] TASS, PETER A.: *Effective desynchronization with bipolar double-pulse stimulation*. Phys. Rev. E, 66(3):036226, 2002.

[TMPP07] TAMASEVICIUS, A., G. MYKOLAITIS, V. PYRAGAS and K. PYRA-GAS: *Delayed feedback control of periodic orbits without torsion in nonautonomous chaotic systems: Theory and experiment*. Phys. Rev. E, 76(2):026203, 2007.

[TRW⁺98] TASS, P., M. G. ROSENBLUM, J. WEULE, J. KURTHS, A. PIKOVSKY, J. VOLKMANN, A. SCHNITZLER and H.-J. FREUND: *Detection of n : m Phase Locking from Noisy Data: Application to Magnetoencephalography*. Phys. Rev. Lett., 81(15):3291–3294, 1998.

[TWWR06] TRONCIU, V. Z., H.-J. WUENSCHE, M. WOLFRUM and M. RADZIUNAS: *Semiconductor laser under resonant feedback from a Fabry-Perot: Stability of continuous wave operation.* Phys. Rev. E, 73:046205, 2006.

[UAI⁺08] UCHIDA, ATSUSHI, KAZUYA AMANO, MASAKI INOUE, KUNIHITO HIRANO, SUNAO NAITO, HIROYUKI SOMEYA, ISAO OOWADA, TAKAYUKI KURASHIGE, MASARU SHIKI, SHIGERU YOSHIMORI, KAZUYUKI YOSHIMURA and PETER DAVIS: *Fast physical random bit generation with chaotic semiconductor lasers.* Nature Photonics, 2(12):728–732, 2008.

[UBB⁺04] USHAKOV, O., S. BAUER, O. BROX, H.-J. WÜNSCHE and F. HENNEBERGER: *Self-Organization in Semiconductor Lasers with Ultrashort Optical Feedback.* Phys. Rev. Lett., 92(4):043902, 2004.

[Ush96] USHIO, T.: *Limitation of Delayed Feedback Control in Nonlinear Discrete-Time Systems.* IEEE Trans. Circuits Syst. I, Fundam. Theory Appl., 43(9):815–816, 1996.

[UWH⁺05] USHAKOV, O. V., H.-J. WÜNSCHE, F. HENNEBERGER, I. A. KHOVANOV, L. SCHIMANSKY-GEIER and M. A. ZAKS: *Coherence Resonance Near a Hopf Bifurcation.* Phys. Rev. Lett., 95(12):123903, 2005.

[VFB95] VOHRA, S. T., L. FABINY and F. BUCHHOLTZ: *Suppressed and induced chaos by near resonant perturbation of bifurcations.* Phys. Rev. Lett., 75:65–68, 1995.

[VIV03] VASSILIEV, V.V., S.M. IL?INA and V.L. VELICHANSKY: *Diode laser coupled to a high-Q microcavity via a GRIN lens.* Appl. Phys. B, 76:521, 2003.

[vL10] LOEWENICH, CLEMENS VON: *Zeitverzögerte Rückkopplungskontrolle torsionsfreier periodischer Orbits.* PhD thesis, Technical University Darmstadt, 2010.

[vLBJ04] LOEWENICH, CLEMENS VON, HARTMUT BENNER and WOLFRAM JUST: *Experimental Relevance of Global Properties of Time-Delayed Feedback Control.* Phys. Rev. Lett., 93(17):174101, 2004.

[vLBJ10] LOEWENICH, CLEMENS VON, HARTMUT BENNER and WOLFRAM JUST: *Experimental verification of Pyragas-Schöll-Fiedler control.* Phys. Rev. E, 82(3):036204, 2010.

[VR98] VANWIGGEREN, GREGORY D. and RAJARSHI ROY: *Communication with Chaotic Lasers.* Science, 279(5354):1198–1200, 1998.

[WBK⁺05] WUENSCHE, H.-J., S. BAUER, J. KREISSL, O. USHAKOV, N. KORNEYEV, F. HENNEBERGER, E. WILLE, H. ERZGRAEBER, M. PEIL, W. ELSAESSER and I. FISCHER: *Synchronization of delay-coupled oscillators: A study on semiconductor lasers.* Phys. Rev. Lett., 94:163901, 2005.

[WH91] WIEMAN, C. E. and L. HOLLBERG: *Using diode lasers for atomic physics.* Rev. Sci. Instrum., 62(1):1, 1991.

[Wie85]   WIESENFELD, K.: *Virtual Hopf phenomenon: A new precursor of period-doubling bifurcations*. Phys. Rev. A, 32(3):1744–1751, 1985.

[WSH08]   WÜNSCHE, H.-J., S. SCHIKORA and F. HENNEBERGER: *Chapter 21 in: Handbook of Chaos Control*. Wiley-VCH, Weinheim, 2nd edition, 2008.

[WYP95]   WIGEN, P.E., M. YE and D.W. PETERMAN: *Controlling chaos in ferromagnetic resonance*. Journ. Magn. Magn. Mat., 140-144:2074–2076, 1995.

[YK88]   YASAKA, HIROSHI and HITOSHI KAWAGUCHI: *Linewidth reduction and optical frequency stabilization of a distributed feedback laser by incoherent optical negative feedback*. Appl. Phys. Lett., 53(15):1360–1362, 1988.

[YWHS06]   YANCHUK, SERHIY, MATTHIAS WOLFRUM, PHILIPP HOVEL and ECKEHARD SCHÖLL: *Control of unstable steady states by long delay feedback*. Phys. Rev. E, 74(2):026201, 2006.

[ZSFYMC94]   ZHANG, L. M., M. NOWELL S. F. YU, D. D. MARCENAC and J. E. CARROLL: *Dynamic analysis of radiation and side mode suppression in second-order DFB lasers using time-domain large signal traveling wave model*. IEEE J. Quant. Electron., 30:1389–1395, 1994.

[ZVA$^+$10]   ZAKHAROVA, A., T. VADIVASOVA, V. ANISHCHENKO, A. KOSESKA and J. KURTHS: *Stochastic bifurcations and coherencelike resonance in a self-sustained bistable noisy oscillator*. Phys. Rev. E, 81(1):011106, 2010.

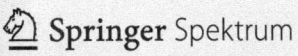